The Logic of Plurality

The Logic of Plurality

J. E. J. ALTHAM

METHUEN & CO LTD
11 NEW FETTER LANE · LONDON EC4

First published 1971
by Methuen & Co. Ltd
11 *New Fetter Lane, London EC4*
© 1971 *J. E. J. Altham*
Printed in Great Britain by
Willmer Brothers Limited, Birkenhead

SBN 416 65920 9

Distributed in the U.S.A.
by Barnes & Noble Inc.

Contents

Preface

Classical elementary logic studies the properties of certain of the logical particles occurring both in everyday argument and in mathematical discourse. These particles include the truth-functional connectives, such as 'not', 'and', 'or', 'if', 'if and only if', the quantificational particles 'all', 'every', 'some', 'none', 'only', the definite article 'the' (in some of its uses), and the relation of identity 'is the same as'. Modern elementary logic was developed largely to complete the formalization of mathematical reasoning, by providing a language-form within which the propositions of mathematics could all be expressed. The logical particles studied by classical logic therefore include all those necessary for the expression of the bulk of mathematical reasoning. It is because of its success in providing the underlying logic of mathematics that this logic is rightly called classical. But the particles it studies do not exhaust all the particles of logical interest, and logicians have constructed various systems that extend classical logic by taking into account some of these other particles. In particular, the intensional connectives 'necessarily', 'possibly' and 'contingently' have been extensively investigated; there is much current work on tense-connectives, such as 'it was the case that' and 'it will be the case that'; other logicians are studying other connectives, such as 'it is known that . . .' and 'it is believed that . . .'. But comparatively little attention has so far been paid to the *quantificational* particles that fall outside the scope of existing classical logic. The present work fills part of this gap.

Among the quantificational particles not studied in classical logic are the phrases 'many', 'few' and 'nearly all', and no satisfactory systematization and investigation of their properties has hitherto been produced. This is somewhat surprising, since the particles are fairly obvious objects of study, their logic is extensional (and hence their study escapes all the difficulties associated with modal logics, particularly quantified modal logics), and systems can be developed for them, which are not markedly more complicated, either formally or semantically, than the systems of classical logic. One reason for the neglect of these quantifiers may be their lack of mathematical importance. Another reason may be that they are thought to be essentially vague concepts. But not all logical interest derives from mathematics, and such vagueness as the concepts possess is, as it turns out, no bar to the development of an exact logic for them.

In this work three systems for the logic of these quantifiers are presented, and it seems convenient to call them systems for the *logic of plurality*. All are systems of natural deduction, and all include the classical predicate calculus.

The basic semantic notion that underlies every system is that of a *many-membered set*, also known as a *manifold*. A manifold is simply a set with many members. In every application of the logic, it is necessary to fix upon a number n, which is to be the least number such that a set of n elements constitutes a manifold. Having fixed upon such a number, which must always be greater than one, we in fact define a manifold as any set containing at least n distinct elements. Two new quantifiers are introduced to the language of predicate calculus: $(Mv)A(v)$, to be read 'for many v, $A(v)$', and $(Nv)A(v)$, to be read 'for nearly all v, $A(v)$'. An interpretation I is said to satisfy $(Mv)A(v)$ if and only if there are n distinct interpretations which differ from I at most in what they assign to a term t, all of which satisfy $A(t)$. And I is said to satisfy $(Nv)A(v)$ if and only if in every n distinct interpretations that differ from I at most in what they assign to t there is at least one that satisfies $A(t)$.

The rules are framed in accordance with these semantic ideas. In considering the interpretation of well-formed formulae of the logic of plurality, the question arises whether we should consider only domains of discourse that contain many members, whether we should consider all non-empty domains, as in the case of classical logic, or whether we should make some third assumption. Each decision leads to a distinct logic. One of the systems presented here yields principles of inference that are valid in every non-empty domain. Another yields principles that can be guaranteed valid only in many-membered domains. In the latter system, the sequents

$$(\forall x)F(x) \vdash (Mx)F(x)$$

$$(Nx)F(x) \vdash (\exists x)F(x)$$

are both provable. In the former system they are not, and, if domains with only a few members are allowed, examples are easily constructible which show the invalidity of these sequents. A third system relies on the assumption that the domain of discourse contains at least $2n - 1$ members, where n is the least number of elements that constitute a manifold. In this system the sequent

$$(Nx)Fx \vdash (Mx)Fx$$

is provable, but not provable in either of the other two. These three systems are set out in their basic form in the second and third chapters, and subjected to certain refinements in the fourth chapter.

The natural deduction system for classical predicate calculus which is used here is very similar to that presented in E. J. Lemmon's book *Beginning Logic* (London, Nelson, 1965), and Lemmon's method of conducting derivations is followed here. This system is not particularly elegant; it was chosen in preference to other possibilities because it is easy to work with, and is familiar to many students who are otherwise not very experienced in logic. Another reason is that Benson Mates uses a very similar system in his *Elementary Logic* (London, O.U.P., 1965), an excellent textbook at a somewhat more advanced level than Lemmon. Discussion of the metatheory of the logic of plurality follows and adapts Mates's discussion.

I should like to express here my general indebtedness to the textbooks of these two authors. An expression of gratitude is also due to Professor Noam Chomsky. In a sparkling lecture he gave in Cambridge (England), he used as examples some sentences involving 'many', and commented that the logic of this quantifier was as yet ill developed. His use of the examples put me straightaway on the trail that led to the systems of this monograph.

CHAPTER ONE

Analysis of plurality-quantifiers

A **The interdefinability of 'many', 'few' and 'nearly all'**
The negation of the proposition
(1) Many men are lovers
is, quite simply
(2) Not many men are lovers.
(2) is equivalent to
(3) Few men are lovers.
(3) in turn is equivalent to
(4) Nearly all men are not lovers.
The negation of (3), namely,
(5) It is not the case that few men are lovers, can also, more idiomatically, be written as
(6) Not a few men are lovers.
(6) is equivalent to (1), and also to
(7) Not nearly all men are not lovers.
Finally,
(8) Not nearly all men are lovers
is equivalent to
(9) Many men are not lovers
and also to
(10) Not a few men are not lovers.
From these and other examples it can be seen that 'many', 'few' and 'nearly all' are related as follows:

$$\text{Many} = \text{not few} = \text{not nearly all not}$$
$$\text{Not many} = \text{few} = \text{nearly all not}$$
$$\text{Not many not} = \text{few not} = \text{nearly all.}$$

Hence 'few' and 'nearly all' can be defined in terms of 'many' and 'not'. Indeed any two of 'many', 'few' and 'nearly all' can be defined in terms of the third and negation.

B   Elements of convention in setting up these interrelations
The proposition (1) is clearly equivalent to
(11) There are many men who are lovers.

From (11), it is obvious that there follows
(12) There is at least one man who is a lover.

(11) thus carries an existential commitment. Since (2) is the negation of (1), (2) is true if and only if (1) is not true. If (12) is not true, then (1) is not true, and so (2) is true. But (2) is equivalent to (3). So if no man is a lover, few men are lovers. Hence 'few' does not carry an existential commitment. This may seem strange, since it is natural to say that few *F*s are *G*s only where there is some *F* which is a *G*. It may be strange, but it is not incorrect. First, there seems to be a distinction between 'few' and 'a few'. The proposition
(13) A few men are lovers
does entail (12), and hence carries existential commitment. A thought that (3) entails (12) may come from confusing (3) with (13). Second, there is intuitively no *contradiction* involved in saying 'Few men love Dorothea—in fact none do'. It is an odd utterance, but its oddity comes from the speaker's saying one thing, and then immediately strengthening it. He starts off by being less informative than he could be, and in the next breath becomes more informative. The utterance is odd in the same way as 'At least one undergraduate obtained a First—in fact three did'. Third, the distinction between (3) and (13) seems to be that between
(14) At most a few men are lovers
    and
(15) At least and at most a few men are lovers.

(14) does not have existential commitment. A reader who still has doubts about whether (3) entails (12) is invited to read (3) in the sense of (14).

'Many men are not lovers' entails that not all men are lovers. Hence 'All men are lovers' entails that not many men are not lovers, i.e. that nearly all men are lovers. So 'Nearly all men are lovers' does not entail that there is any man who is not a lover. Three points similar to those made about 'few' can be made to anybody who doubts the correctness of this result. First, there seems to be a distinction between 'Nearly all men are lovers' and

(16) All but a few men are lovers.

Second, there is no contradiction in saying 'Nearly everybody loves Dorothea – in fact everybody does'. There is an oddity in this, but it comes from the immediate passage from a weaker to a stronger statement. Third, the distinction between 'Nearly all men are lovers' and (16) seems to be that between

(17) At least all but a few men are lovers

   and

(18) At least and at most all but a few men are lovers.

A reader who still has doubts over the correctness of these points is invited to read 'Nearly all' as 'At least nearly all'.

According to the conventions of classical logic, if there are no Fs, then all Fs are Gs. But, as has already been pointed out, if all Fs are Gs, then nearly all Fs are Gs. The conventions of classical logic are adopted here. So if there are no Fs, nearly all Fs are Gs. So 'nearly all Fs are Gs' can be *vacuously* true. It can, however, be vacuously – or semi-vacuously – true in another way. If there are not many Fs, it follows that there are not many Fs which are not Gs. But then nearly all Fs are Gs. This result should occasion neither more nor less surprise than the corresponding result in classical predicate logic, that if nothing is F, everything that is F is G. Further consequences are that from 'Nearly all Fs are Gs' it does not follow either that there is an F that is G, or that there are many Fs that are G.

If the interdefinabilities of Section **A** are genuine, it follows that 'Many Fs are G' and 'Nearly all Fs are not G' cannot both be true. Here an objection may come to mind. For surely, one might say, nearly everybody is not colour-blind, but there are yet many people who are colour-blind. Surely there are many severe schizophrenics, but nearly everybody is not severely schizophrenic. Surely although nearly all prime numbers are greater than 10^{100}, there are many prime numbers which are not greater than 10^{100}. It is not clear, however, that these examples are convincing. There seem to be two conflicting tendencies in the casual use of these quantifiers. One tendency is to say that there are many where there are just a lot, in some vague sense, and to say that nearly all have some property when an overwhelmingly high proportion have some property. According to these uses, many things in some very large class may have some property, while nearly everything in that class does not have that property. This contradicts the interdefinability. But there is another tendency in the use of these quantifiers, which makes what counts as nearly all relative to what counts as many, and supports the interdefinitions. The supposition of these two tendencies seems plausible, and seems to be the only way of accounting for the simultaneous plausibility and yet oddity of the above examples.

In adopting the interdefinitions there is perhaps an element of stipulation that just one of these tendencies is to be consistently followed. There may also be some stipulation in the points that have been made about existential commitments. Such stipulation is not damaging to the interest of the logic that results from it, if (1) the resulting logic is workable and reasonably rich in formal principles, and (2) the stipulations have a substantial basis in informal usage. The conventions adopted here meet condition (2); they are entirely comparable to the procedure whereby a truth-functional connective, the so-called material conditional, is used in logic as the closest representative of the informal 'if'. Whether the resulting logic is reasonably rich can only be judged from the sequel.

C   The analogy with the relations between 'some', 'none' and 'all'

The interrelations of 'many', 'few' and 'nearly all' that have been set up, together with the conventions adopted, constitute a substantial analogy between these quantifiers and the classical quantifiers. Just as any two of 'some', 'none' and 'all' may be defined in terms of the third together with negation, so any two of 'many', 'few' and 'nearly all' may be defined in terms of the third together with negation. In this analogy, 'some' corresponds to 'many', 'none' to 'few', and 'all' to 'nearly all'. 'Many' is a stronger existential quantifier than 'some'; 'few', like 'none', is non-existential, but is a weaker quantifier; 'nearly all', like 'all', may be vacuously satisfied, and is also a weaker quantifier.

The analogy extends so far that as a first shot one might conjecture that any valid principle of classical logic becomes a valid principle of the logic of plurality if the existential quantifier is replaced by a quantifier for 'many', and the universal quantifier by a quantifier for 'nearly all'. This conjecture, however, is easily refuted. There are indeed many principles that remain valid upon such transformation, and some of them have been mentioned already. They include such principles as

(19) $\sim(\exists x)Fx \rightarrow (\forall x)(Fx \rightarrow Gx)$

   which becomes

(20) If not many things are F, then nearly everything which is F is G

   and

(21) $(\exists x)(Fx \,\&\, Gx) \rightarrow (\exists x)Fx \,\&\, (\exists x)Gx$

   which becomes

(22) If many things are both F and G, then many things are F, and many things are G.

   Many other such principles, whose transformation in the way suggested are valid principles of the logic of plurality, are easily discoverable. But other valid principles of classical logic become invalid under this transformation. For instance

(23) $(\forall x)Fx \,\&\, (\forall x)Gx \rightarrow (\forall x)(Fx \,\&\, Gx)$.

   This becomes

(24) If nearly everything is F, and nearly everything is G, then nearly everything is both F and G,

   which is not a valid principle.

   Similarly,

(25) $(\exists x)(Fx \vee Gx) \rightarrow (\exists x)Fx \vee (\exists x)Gx$

is a valid principle of classical logic. Its transformation, however, is not valid. It is

(26) If many things are either F or G, then many things are F or many things are G.

   It is also shown later that the analogue of

(27) $(\forall x)Fx \rightarrow (\exists x)Fx$

   namely

(28) If nearly everything is F, then many things are F

is also not valid. This last example highlights the reason for the limits to the analogy, that 'many' is stronger than 'some', while 'nearly all' is weaker than 'all'. One can think up many other invalid analogues of valid principles of classical logic. One of the more interesting is

(29) If nearly everything is F, and nearly everything that is F is G, then nearly everything is G.

   A counter-example to this principle is given later.

   The most interesting differences between classical and

plurality quantifiers, however, appear only when multiple quantifications are considered. Analysis of the reasons for the differences here leads directly to a suitable logic of plurality. This is the next task.

D   Limits to the analogy in cases of multiple quantification

Consider the proposition

(30) Some boy loves some girl.

Speaking in the jargon of logic, the first 'some' in (30) includes the second within its scope. If (30) is transformed into the passive, however, as

(31) Some girl is loved by some boy

the scope of 'some girl' comes to include that of 'some boy'. That is to say, the order of the existential quantifiers in the prefix is reversed. (30) can be expressed as

(32) For some boy, for some girl, that boy loves that girl.

But (31) would analogously be expressed as

(33) For some girl, for some boy, that girl is loved by that boy.

In spite of the reversal of the order of the quantifiers in the prefix, (30) and (31) are logically equivalent. There is in fact a general law of logic according to which a prefix composed exclusively of existential quantifiers may be permuted at will without change of logical force.

Now compare

(34) Many boys love many girls.

Transformed into the passive in the natural way, (34) becomes

(35) Many girls are loved by many boys.

But now a little reflection reveals that (34) and (35) *are not logically equivalent*. A small model makes this clear. Let us suppose that there are five boys, Tom, Dick, Harry, Bill and Fred, and five girls, Jane, Sarah, Bess, Meg and Angela. Let us suppose that a necessary and sufficient condition that a group of boys or girls be a group of *many* boys, or many girls, is that it should contain at least three members. Finally, suppose the boys' loves are distributed as follows: Tom loves Jane, Sarah and Bess; Dick loves Bess, Meg and Angela; Fred loves Jane, Meg and Angela; Harry and Bill love no girl. The loves may be put in a table, where the arrow represents the relation of love.

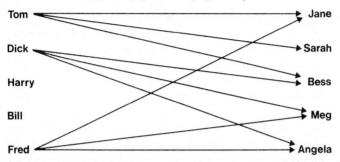

Under our assumptions, (34) is true, because each of a group of three boys loves three girls. But (35) is false, since there is no group of three girls, each of whom is loved by three boys. A similar model might show the possibility that (35) be true while (34) were false. (34) and (35) are therefore logically independent propositions. Consequently, the interchange of two 'manys' in a prefix consisting exclusively of 'manys' alters the logical force of what is expressed. The order of these quantifiers matters.

In a similar way it is possible to show that

(36) Nearly all boys love nearly all girls

neither entails, nor is entailed by

(37) Nearly all girls are loved by nearly all boys.

But if a prefix consists exclusively of universal quantifiers, they can be permuted at will without change of logical force. For instance,

(38) Every boy loves every girl

is logically equivalent to

(39) Every girl is loved by every boy.

These non-equivalences are of the utmost importance. To gain an understanding of them, and the difference from the classical quantifiers, is the key to the construction of a logic of plurality.

E Analysis in terms of the concept of a manifold

In explaining how the small model verified (34) while falsifying (35), appeal was made to the notion of a *set's having many members*, and, to make the matter precise, it was stipulated that for a set to have many members (in the model), it must have at least three. This suggests that a natural way to analyse sentences involving plural quantifications is in terms of notions belonging to set theory. To this end, the concept of a *manifold* is first defined. A manifold is simply a set that contains many members. In any context, a certain number n is fixed upon as being the least number of objects that can form a manifold. Having done that, a set is defined to be a manifold if and only if it contains at least n members. Within certain limits, the number n is arbitrary, but in any context n must be greater than one, since it is certain that there is no context in which one would say that there are many Fs if there were only one F.

Now reconsider

(1) Many men are lovers.

In terms of the concept of a manifold, this can be analysed as

(40) There is a manifold of men, all of whom are lovers.

Or, more fully

(41) There is a set, which is a manifold, all of whose members are men, and every object in that set is a lover.

Let us introduce the symbol 'M', to be read 'is a manifold'. Then (41) can be symbolically transcribed as

(42) $(\exists x)(Mx \,\&\, (\forall y)(y \in x \to y$ is a man$) \,\&\, (\forall z)(z \in x \to z$ is a lover$))$.

More simply,

(43) $(\exists x)(Mx \,\&\, (\forall y)(y \in x \to y$ is a man and y is a lover$))$.

Sentences involving 'few' or 'nearly all' can also be expressed in this symbolism. For instance,

(3) Few men are lovers

is simply the negation of (43), and this can be transformed into

(44) $(\forall x)(Mx \to (\exists y)(y \in x \,\&\, (y$ is not a man $\vee y$ is not a lover$)))$.

i.e. to say that few men are lovers is to say that in any manifold whatever there is some element that is either not a man or not a lover. This seems to give just the meaning of (3). 'Nearly all' can clearly be treated by the same means.

(34) and (35) are more complicated, but are equally amenable to transcription into this symbolism.

(34) Many boys love many girls.

(34) says

(45) There is a manifold of boys, and for each boy in that manifold there is a manifold of girls such that that boy loves every girl in that manifold.

(35), on the other hand, says

(46) There is a manifold of girls, and for every girl in that manifold there is a manifold of boys such that every boy in that manifold loves that girl.

(45) and (46) are transcribable respectively as

(47) $(\exists x)\{Mx \ \& \ (\forall y)(y \in x \rightarrow y$ is a boy $\&$
$(\exists z)[Mz \ \& \ (\forall w)(w \in z \rightarrow w$ is a girl $\& \ y$ loves $w)])\}$.

(48) $(\exists z)\{Mz \ \& \ (\forall w)(w \in z \rightarrow w$ is a girl $\&$
$(\exists x)[Mx \ \& \ (\forall y)(y \in x \rightarrow y$ is a boy $\& \ y$ loves $w)])\}$.

Upon reflection, it becomes clear that (47) and (48) are not equivalent. This symbolism seems to be adequate to reveal the logical structure of plural quantifications. One point of interest that emerges from it is that a quantification of the form 'many' is revealingly represented as a *double* quantification. One of the two quantifiers involved in it is existential, the other universal. This symbolism is, however, rather cumbersome, and it would be better to find a similarly revealing symbolism within the framework of predicate logic, i.e. a *first-order* symbolization. This can be obtained from the set-theoretic symbolization in a straightforward way.

F Double-quantifier form of first-order symbolization

(1) was transcribed as

(43) $(\exists x)(Mx \ \& \ (\forall y)(y \in x \rightarrow y$ is a man $\& \ y$ is a lover)).

This sentence expresses a proposition, which asserts the existence of a manifold and ascribes a (complex) property to it. A first-order sentence that does the same job is obtainable as follows: a new quantifier is introduced, '$(\exists Mx)$', to be read 'there is a manifold of xs . . .'. A sentence beginning with this quantifier will proceed to speak of all the objects in this manifold. So a subscript is placed under the subsequent universal quantifier, to show that the range of that quantifier is restricted to the objects in that manifold, thus: $(\forall y_x)$. Now (1) can be transcribed as

(49) $(\exists Mx)(\forall y_x)(y$ is a man $\& \ y$ is a lover).

The effect of restricting the universal quantifier is the same as the effect of the antecedent of the conditional in (43). (49) therefore expresses what (43) expresses in a different form. A further simplification can be obtained by using a special variable 'm', which ranges only over men, in place of 'x' in (49). The effect of this device is the same as that of the first conjunct in (48). (49) may therefore be abbreviated to

(50) $(\exists Mm)(\forall y_m)(y$ is a lover).

Similarly, we can transcribe (34) and (35) as (respectively)

(51) $(\exists Mb)(\forall y_b)(\exists Mg)(\forall z_g)(y$ loves $z)$

(52) $(\exists Mg)(\forall z_g)(\exists Mb)(\forall y_b)(y$ loves $z)$

where 'b', and 'g' range only over boys and girls respectively.

Now the non-equivalence of (51) and (52) is strictly comparable to the situation in predicate logic, where

(53) $(\exists b)(\forall y)(\exists g)(\forall z)Fbygz$

is not equivalent to the result of switching the second pair of quantifiers with the first, i.e.

(54) $(\exists g)(\forall z)(\exists b)(\forall y)Fbygz$.

There are other respects also in which '$(\exists Mx)$' behaves like an existential quantifier. For instance, a further new quantifier '$(\forall Mx)$', to be read 'For every manifold of xs . . .', may be introduced. Now reconsider

(4) Nearly all men are not lovers.

This means the same as

(55) In every manifold of men there is at least one who is not a lover.

(55) may be transcribed, using the new quantifier, and a restricted variable, as

(56) $(\forall Mm)(\exists y_m)\sim(y$ is a lover).

Now (4) is equivalent to

(2) Not many men are lovers.

(2) may be transcribed as the negation of (50), i.e.

(57) $\sim(\exists Mm)(\forall y_m)(y$ is a lover).

(57) is therefore equivalent to (56). Consequently, just as in predicate logic quantifiers may be interchanged so that

(58) $(\forall x)(\exists y)\sim A$

is equivalent to

(59) $\sim(\exists x)(\forall y)A$

so, in the logic of plurality '$(\forall Mx)(\exists y_x)\sim A$' can be transformed into '$\sim(\exists Mx)(\forall y_x)A$'.

Similarly

(60) Not many boys love many girls

comes out as

(61) $(\forall Mb)(\exists y_b)(\forall Mg)(\exists z_g)\sim(y$ loves $z)$

i.e. in every manifold of boys there is at least one such that in every manifold of girls there is at least one whom he does not love – a complicated, but genuine equivalent of (60).

This form of symbolization is philosophically revealing. It shows how 'many', 'few' and 'nearly all' may be exhibited as each involving a double quantification, of a kind which reveals its difference from 'some', 'none' and 'all' respectively. At the same time it shows how the interrelations between the plural quantifiers are explicable in terms of the interrelations between the classical quantifiers. Further, the present symbolism has a natural semantics, similar to that for predicate logic. To explain the semantics, it is assumed that plural quantifiers always occur in one of the forms '$(\forall Mv)(\exists w_v)$' or '$(\exists Mv)(\forall w_v)$'.

In predicate logic, '$(\forall v)(\exists w)A$' is true under an interpretation I if and only if for every element in the domain of discourse there is an element such that A. And '$(\exists v)(\forall w)A$' is true if and only if there is an element such that A for all elements in the domain. Correspondingly, '$(\forall Mv)(\exists w_v)A$' is true if and only if for every manifold M in the domain, '$(\exists w)A$' is true where the values of 'w' are restricted to the elements of M. And '$(\exists Mv)(\forall w_v)A$' is true if and only if there is a manifold in the domain such that '$(\forall w)A$' is true where the values of 'w' are restricted to the elements of M.

Now remember that a manifold is any set with n members. Consequently, '$(\forall Mv)(\exists w_v)A$' is true if and only if in every set of n distinct elements there is at least one such that A, and '$(\exists Mv)(\forall w_v)A$' is true if and only if there are n distinct elements, every one of which is such that A.

Consider in this connection

(62) $(\exists Mx)(\forall y_x)Fy$.

This is true if and only if there are n elements, $a_1, a_2, \ldots a_n$, such that 'Fa_1', 'Fa_2', $\ldots$ 'Fa_n' are all true.

Now consider

(63) $(\exists Mx)(\forall y_x)(\exists Mz)(\forall t_z)Ryt$.

This is true if and only if, for some $a_1, \ldots a_n$, all of

$(\exists Mz)(\forall t_z)Ra_1t$
$(\exists Mz)(\forall t_z)Ra_2t$
. . . .
. . . .
$(\exists Mz)(\forall t_z)Ra_nt$

are true. These are all true if and only if, for every $a_i(1 \leq i \leq n)$, there are n elements $b_{i1}, \ldots b_{in}$, such that all of

$$Ra_i b_{i1}, \ldots Ra_i b_{in}$$

are true. So a formula of the form (63) is true if and only if elements can be found such that all of an $n \times n$ array of the following form are true:

$$Ra_1 b_{11}, \; Ra_1 b_{12}, \ldots Ra_1 b_{1n}$$
$$Ra_2 b_{21}, \; Ra_2 b_{22}, \ldots Ra_2 b_{2n}$$
$$\cdots\cdots\cdots\cdots\cdots\cdots$$
$$\cdots\cdots\cdots\cdots\cdots\cdots$$
$$Ra_n b_{n1}, \; Ra_n b_{n2}, \ldots Ra_n b_{nn}$$

On the other hand, a formula of the form

(64) $(\exists Mz)(\forall t_z)(\exists Mx)(\forall y_x)Ryt$

is true if and only if elements can be found such that all of an array of *this* form are true

$$Ra_{11} b_1, \; Ra_{12} b_1, \ldots Ra_{1n} b_1$$
$$Ra_{21} b_2, \; Ra_{22} b_2, \ldots Ra_{2n} b_2$$
$$\cdots\cdots\cdots\cdots\cdots\cdots$$
$$\cdots\cdots\cdots\cdots\cdots\cdots$$
$$Ra_{n1} b_n, \; Ra_{n2} b_n, \ldots Ra_{nn} b_n$$

Clearly, an array of the first form may be true without any array of the second form being true, and vice versa. The semantic interpretation thus explains the non-equivalence of (63) and (64). In the small model given earlier, a 3×3 array of the first form can be given, but no 3×3 array of the second form can be given, all of whose elements are true.

This method of symbolizing plural quantifications has been given because it is philosophically revealing. It is not, however, the best method to adopt for a system of the logic of plurality. It remains long-winded and cumbrous, though less so than the set-theoretic symbolism. Instead, it is better to adopt a *single-quantifier* symbolic representation, and ensure that the features that made the double-quantifier form revealing are preserved in the rules of valid inference. This will now be done.

G Single-quantifier form of first-order symbolization: semantics for this form

Two new quantifiers are introduced, '(Mx)', to be read 'For many $x \ldots$', and '(Nx)' to be read 'For nearly all x'.

(1) is now written very simply as

(65) $(Mx)(x$ is a man $\&$ x is a lover).

'Nearly all men are lovers' means 'for nearly every x, if x is a man, then x is a lover'. It is written

(66) $(Nx)(x$ is a man $\rightarrow x$ is a lover).

The rules of truth for formulae containing these quantifiers are as follows. It is assumed that some number n has been decided upon as the least number of elements that form a manifold.

$(Mv)A(v)$ is true under I if and only if $A(v)$ is true (under I) for at least n distinct values of v.

$(Nv)A(v)$ is true under I if and only if $A(v)$ is true for at least one of every n distinct values of v.

With these rules it can be seen, as before, that two 'manys', or two 'nearly alls', cannot be permuted without change of logical force, i.e.

(67) $(Mx)(My)Fxy$ is not equivalent to $(My)(Mx)Fxy$

   and

(68) $(Nx)(Ny)Fxy$ is not equivalent to $(Ny)(Nx)Fxy$.

Also, it is clear that the rules justify the definition of one quantifier in terms of the other. Either $(Mv)A(v)$ may be defined as $\sim(Nv)\sim A(v)$, or $(Nv)A(v)$ may be defined as $\sim(Mv)\sim A(v)$.

The rules of truth give rise to arrays precisely similar to those that the semantics of Section **F** gave rise to. The gain in simplicity amply compensates for the loss of the clarity gained by the double-quantifier representation.

The rules of truth give rise to a problem. $(Nv)A(v)$ is to be true if and only if for every n distinct values of v, at least one makes $A(v)$ come out true. But suppose there are no n distinct values of v, because the universe of discourse contains too few members. Is $(Nv)A(v)$ to be vacuously true, or is it to be assumed that only domains with at least n members are to be admitted? This work provides no answer to this question. Instead, *two* systems are developed. One, the weaker, contains those principles that are valid in every non-empty domain, and $(Nv)A(v)$ may be vacuously true. Domains for this system do not have to contain many members, but must contain at least one. The second, stronger system contains those principles that are valid in every domain with at least n members. $(Nv)A(v)$ cannot then be vacuously true. The stronger system is perhaps more natural, but the weaker system has the advantage that it makes the same assumption about domains as the classical predicate logic. It may for all that be considered merely a curiosity, but at least it is an ingenious curiosity.

(It should be noticed that the vacuity discussed here is not the same as the vacuous truth of '$(Nx)(Fx \rightarrow Gx)$' where '$\sim(Mx)Fx$'. This latter is common to both systems.)

CHAPTER TWO

The weaker system W

Introduction

This chapter contains an exposition of the logic of plurality by the method of natural deduction. A logic constructed by this method proceeds by laying down a set of basic rules of inference for the notions it treats, which permits the derivation of certain formulae as conclusions, given certain formulae as assumptions. The set of basic rules is so chosen that with its aid a large number of principles may be derived of the form 'Given assumptions of such and such forms, a conclusion of such and such a form may be deduced'. This method has become the most popular one in elementary logic texts, for the two reasons (1) that proofs tend to be easier to discover than by other methods, and (2) that it corresponds most naturally with the role of logic as providing principles whereby conclusions may be validly deduced from assumptions. It is for these reasons, together with the fact that it is now the most familiar method, that it is adopted here.

The procedure adopted is to take some well-known system of classical predicate logic, to add to its vocabulary the new quantifiers '(Mx)' and '(Nx)', and to add to its rules a further four basic rules of inference that essentially involve these quantifiers. Proofs of a fair number of more or less interesting principles are then given, by means of these rules. It is then shown that there are unnecessarily many of these rules, since some can be derived from others. An exact statement is then given of how the rules are to be interpreted, and the two crucial matters of the soundness and completeness of the system are discussed. A system is sound if it is possible to prove only valid principles in it; it is complete if it is possible to prove *every* valid principle in it.

The system of classical predicate logic used is almost identical with that given in E. J. Lemmon's book *Beginning Logic*. A very similar system occurs in Benson Mates's book *Elementary Logic*. Classical predicate logic is merely sketched here, and for further information on it the reader is referred to these two books.

A   The vocabulary and rules of well-formedness

The vocabulary of the system W consists of the following symbols:

(1) The proper names (or *terms*)

$$a, b, c, d, a_1, b_1, \ldots a_n, b_n, \ldots$$

(2) The variables

$$x, y, z, x_1, y_1, \ldots x_n, y_n, \ldots$$

(3) The predicate letters of degree n, for each $n \geq 0$

$$F^0, G^0, H^0, F_1{}^0, G_1{}^0, \ldots F_n{}^0, G_n{}^0, \ldots$$
$$F^1, G^1, H^1, F_1{}^1, G_1{}^1, \ldots F_n{}^1, G_n{}^1, \ldots$$
$$F^2, G^2, \ldots$$
$$\cdots \cdots \cdots$$
$$\cdots \cdots \cdots$$
$$F^n, G^n, \ldots$$
$$\cdots \cdots \cdots$$

(4) The predicate constant of degree 2, identity

$$=$$

(5) The connectives

    $\sim$ (not)

    $\&$ (and)

    $\vee$ (or)

    $\rightarrow$ (if)

(6) The quantifier-signs

    $\exists$ (existential)

    $\forall$ (universal)

    M (plural)

    N (penuniversal)

(The word 'penuniversal' is devised on the model of 'penultimate' or 'penumbra', from the Latin 'paene' (almost) and 'universalis'.)

(7) The brackets

    (,)

Predicate letters of degree 0 are known as *propositional variables*, and it is convenient to use the letters '*P*', '*Q*', '*R*', etc., for these.

A *formula* of the system *W* is any finite sequence of symbols from the lists 1–7. In order to define the concept of well-formed formula (wff), and for other purposes too, some syntactic variables are needed.

'*t*', '*t₁*', '*t₂*', . . .   are to be syntactic variables ranging over terms.

'*K*', '*K₁*', '*K₂*', . . .   are to be syntactic variables ranging over predicate letters.

'*A*', '*B*', '*C*', '*D*', . . .   are to be syntactic variables ranging over wffs.

'*v*', '*w*', '*v₁*', . . .   are to be syntactic variables ranging over variables.

The expressions '$A(t)$', '$B(v)$', and others like them, are used to refer to a formula containing a term t, or a formula containing a variable v.

Now let K be a predicate letter of degree n, and let $t_1, \ldots t_n$ be terms (not necessarily distinct). Then $Kt_1, \ldots t_n$ is an *atomic sentence*. n may be zero, so propositional variables are atomic sentences. Similarly, let t_1 and t_2 be terms. Then $t_1 = t_2$ is an atomic sentence. Thus an atomic sentence is any formula which is either a propositional variable, or a predicate letter of degree n followed immediately by a sequence of n terms, or the identity-sign flanked with a term on both sides. The well-formed formulae of the calculus of plurality can now be defined.

(1) Any atomic sentence is a wff.

(2) If A is a wff, then $\sim A$ is a wff.

(3) If A and B are wffs, then $(A \ \& \ B)$, $(A \lor B)$, and $(A \to B)$ are wffs.

(4) Let $A(t)$ be a wff containing a term t, and let v be a variable not occurring in $A(t)$. Let $A(v)$ be a formula resulting from $A(t)$ by replacing at least one occurrence of t in $A(t)$ by v. Then $(\forall v)A(v)$, $(\exists v)A(v)$, $(Mv)A(v)$, and $(Nv)A(v)$ are wffs.

(5) If a formula is not a wff in virtue of 1–4, then it is not a wff. If $A_1, A_2, \ldots A_n, B$ are wffs, then

$$A_1, A_2, \ldots A_n \vdash B$$

is a *sequent-expression*.

The Rule 4 is more complicated than it might be. Its point, when taken with the other rules, is to ensure (1) that no wff contains any free variable, so that all wffs are *sentences*, and (2) that no quantifier occurs without there being a variable in the succeeding expression that is bound by it. At the cost of a little complexity in the rules of formation, satisfaction of these two points makes for naturalness in the system. For further information, see Lemmon's book *Beginning Logic*.

Again following Lemmon, a formula A is defined to be a *propositional function* in the n variables $v_1, \ldots v_n$ if and only if $(\forall v_1)(\forall v_2), \ldots (\forall v_n)A$ is a wff. The case $n = 0$ is allowed.

It is convenient to adopt certain conventions of definitional abbreviation. First, $(A \leftrightarrow B)$ is introduced as an abbreviation for $((A \to B) \,\&\, (B \to A))$. Second, the outermost pair of brackets in a wff may be dropped. Thus instead of $(A \,\&\, B)$, we may write simply $A \,\&\, B$. Third, we adopt the usual conventions for further omission of brackets by ordering the connectives in strength, so that '$\sim$' binds more closely than ' $\&$ ', which binds more closely than ' $\vee$ ', which binds more closely than '$\to$' or '$\leftrightarrow$'. Thus upon replacement of brackets, the formula

$$\sim\!A \,\&\, B \vee C \to D$$

becomes

$$(((\sim\!A \,\&\, B) \vee C) \to D).$$

Fourth, we shall be dealing extensively in what follows with wffs of the following forms:

$$A(t_1) \,\&\, A(t_2) \,\&\, \ldots \,\&\, A(t_n)$$
$$A(t_1) \vee A(t_2) \vee \ldots \vee A(t_n)$$
$$\sim\!(t_1 = t_2) \,\&\, \sim\!(t_1 = t_3) \,\&\, \ldots \,\&\, \sim\!(t_1 = t_n) \,\&\, \sim\!(t_2 = t_3) \,\&\, \ldots$$
$$\&\, \sim\!(t_2 = t_n) \,\&\, \ldots \,\&\, \sim\!(t_{n-1} = t_n).$$

A formula of the first form will be called an n-termed conjunction, one of the second form an n-termed disjunction, and one of the third form an n-termed non-identity. An n-termed non-identity asserts of a set of n objects (not necessarily distinct) that they *are* all distinct.

An n-termed conjunction is abbreviated to

$$\underset{1 \leq i \leq n}{\&} \; A(t_i).$$

An n-termed disjunction is abbreviated to

$$\bigvee_{1 \le i \le n} A(t_i).$$

An n-termed non-identity is abbreviated to

$$(t_i \neq t_j)_{1 \le i,\, j \le n,\, i \neq j}.$$

The subscripts will generally be omitted, where no misunder-standing can result from doing so.

The importance of wffs of these forms will become clear from the rules for the plural and penuniversal quantifiers. In the case of wffs of these forms, it is always assumed that n is *greater* than 1.

B Rules of derivation

A derivation is a finite sequence of wffs, each of which either is an *assumption*, or results from previous members of the sequence by application of one of the *rules of derivation*. If a wff in a derivation is an assumption, it results from the application of the *rule of assumptions* (Rule A). If a wff is the nth wff in a derivation, it is said to occur on the nth *line* of that derivation. Each rule of derivation specifies on what assumptions the wff that results from its application depends. If a wff B occurs on the last line of a derivation, and there depends on the assump-tions $A_1, \ldots A_n$, and only on these, the derivation is said to be a *proof of B from the assumptions $A_1, \ldots A_n$*. It is also said to be a proof of the sequent-expression $A_1, \ldots A_n \vdash B$. A sequent-expression for which there exists a proof is called a *sequent*. If B occurs on the last line of a derivation, and there depends on *no* assumptions, the derivation is said to be simply a *proof of B*. A wff for which there exists a proof is a *theorem*. We write $\vdash B$ to mean that B is a theorem.

Derivations are written as a sequence of lines down the page. If a wff occurs on a line, we write the number of that line in the derivation immediately to the left of that wff. On the far left are written the numbers of the assumptions upon which the wff depends in that derivation. The number of an assumption is the number of the line on which that assumption results by an application of the rule of assumptions. On the far right of a wff are written, first, the numbers of the lines from which that wff results by application of a rule of derivation, and second, the name of that rule of derivation. This is the method of writing out derivations used by Lemmon, to whose book (already cited) the reader is referred for further explanation if need be.

We give the rules of derivation in five groups: (1) the rule of assumptions, (2) the rules for the connectives, (3) the rules for the classical quantifiers, (4) the rules for the plural and pen-universal quantifiers, and (5) the rules for identity.

(1) RULE OF ASSUMPTIONS (A)

Any wff may be entered as a line of a derivation at any stage, depending upon itself. So on the far left we write the number of that line, and on the far right we write A.

(2) RULES FOR THE CONNECTIVES

(i) *Modus ponendo ponens* (MPP)

If wffs A and $A \to B$ occur as existing lines, B may be entered as a further line. B depends on any assumptions upon which either A or $A \to B$ depend. So we write on the far left all numbers that occur on the far left of A on its line, *and* all numbers that occur on the far left of $A \to B$ on *its* line. To the far right we write the numbers of the lines where A and $A \to B$ occur, and the name MPP.

(ii) *Modus tollendo tollens* (MTT)

If $\sim B$ and $A \to B$ occur as existing lines, $\sim A$ may be entered as a further line, depending on any assumptions upon which either $\sim B$ or $A \to B$ depend. We write the numbers of these assumptions on the far left, and on the far right, the numbers of the lines where $\sim B$ and $A \to B$ occur, and MTT.

(iii) *Double negation* (DN)

If A occurs as an existing line, $\sim \sim A$ may be entered as a further line, and vice versa. In either case, the further line depends on the same assumptions as the earlier. To the far right we write DN.

(iv) *Conditional proof* (CP)

If a derivation constitutes a proof of B from assumptions $A_1, \ldots A_n$, A, $A \to B$ may be entered as a further line, depending on assumptions $A_1, \ldots A_n$.

(*iii*) and (*iv*) have been given in a summary form, as will the remaining rules for the connectives with one exception. What in each case must be written on the far left and the far right should be clear from the explanations already given.

(*v*) *&-introduction* (*&I*)
Given A and B as existing lines, $A \And B$ may be entered as a further line, depending on any assumptions on which either A or B depends.

(*vi*) *&-elimination* (*&E*)
Given $A \And B$ as an existing line, A may be entered as a further line. Also, given $A \And B$, B may be entered as a further line. In either case, the further line depends on the same assumptions as $A \And B$.

(*vii*) ∨-*introduction* (∨*I*)
Given A, $A \lor B$ may be entered. Also, given B, $A \lor B$ may be entered. In either case $A \lor B$ depends on the same assumptions as the line whence it results.

(*viii*) ∨-*elimination* (∨*E*)
Given $A \lor B$, depending on assumptions $A_1, \ldots A_n$, and a proof of C from $B_1, \ldots B_m, A$, and a proof of C from $C_1, \ldots C_k, B$, C may be entered as further line, depending on $A_1, \ldots A_n$, $B_1, \ldots B_m, C_1, \ldots C_k$. On the far right we cite

(*a*) The number of the line $A \lor B$.
(*b*) The number of the line where A is assumed.
(*c*) The number of the line where C occurs, derived from $B_1, \ldots B_m, A$.
(*d*) The number of the line where B is assumed.
(*e*) The number of the line where C occurs, derived from $C_1, \ldots C_k, B$.

Any or all of n, m, k may be zero.

The idea behind this rule is simply that if a conclusion follows from each of two wffs, together with given assumptions, then it follows from the disjunction of those two wffs, together with those given assumptions, and any on which the disjunction depends.

(*ix*) *Reductio ad absurdum* (*RAA*)

Given a proof of B & $\sim B$ from $A_1, \ldots A_n$, A, $\sim A$ may be entered, depending on $A_1, \ldots A_n$.

(*x*) The definition of '$\leftrightarrow$' permits the replacement of $A \leftrightarrow B$ by $(A \rightarrow B)$ & $(B \rightarrow A)$, and vice versa, in each case depending on the same assumptions.

(3) THE RULES FOR THE CLASSICAL QUANTIFIERS

(*i*) *Universal elimination* (*UE*)

Let $A(v)$ be a propositional function in v, and let t be a term. Let $A(t)$ be the result of replacing all and only occurrences of v in $A(v)$ by t. Then given $(\forall v)A(v)$, $A(t)$ may be entered as a further line, depending on the same assumptions as $(\forall v)A(v)$.

(*ii*) *Existential introduction* (*EI*)

Let $A(v)$, t and $A(t)$ be as in (*i*). Then given $A(t)$, $(\exists v)A(v)$ may be entered as a further line, depending on the same assumptions as $A(t)$.

(*iii*) *Universal introduction* (*UI*)

Let $A(t)$ be a wff containing the term t, and let v be a variable that does not occur in $A(t)$; let $A(v)$ be the propositional function in v that results from replacing all and only occurrences of t in $A(t)$ by v. Then given $A(t)$, $(\forall v)A(v)$ may be entered as a further line, depending on the same assumptions as $A(t)$, provided that t occurs in no assumption upon which $A(t)$ depends.

(*iv*) *Existential elimination* (*EE*)

Let $A(t)$, v and $A(v)$ be as in (*iii*). Then given a proof of $(\exists v)A(v)$ from assumptions $A_1, \ldots A_n$, and a proof of C from $A(t)$, $B_1, \ldots B_m$, C may be entered as a further line, depending upon $A_1, \ldots A_n$, $B_1, \ldots B_m$, provided that t occurs neither in C nor in any of $B_1, \ldots B_m$.

(The system so far is essentially that of Lemmon's book, except that Lemmon's device of using a special category of *arbitrary names* has been dropped. Arbitrary names are theoretically superfluous, and the distinction Lemmon makes between them and proper names is quite specious, as is shown by the fact that in discussing the *interpretation* of wffs (p. 156), he makes no distinction between the interpretation of proper names and that of arbitrary names, so that arbitrary names are just proper names under a different name.)

(4) THE RULES FOR THE PLURAL AND PENUNIVERSAL QUANTIFIERS

(i) Penuniversal elimination (NE)

Let $A(v)$ be a propositional function in v, and let $t_1, t_2, \ldots t_n$ be n distinct terms (n strictly greater than 1). Let $A(t_1)$, $A(t_2), \ldots A(t_n)$ be the results of replacing all and only occurrences of v in $A(v)$ by $t_1, t_2, \ldots t_n$ respectively. Then given $(Nv)A(v)$, the wff

$$(t_i \neq t_j) \to \bigvee A(t_i)$$

may be entered as a further line, depending upon the same assumptions as $(Nv)A(v)$.

(ii) Plural introduction (MI)

Let $A(v)$, $t_1, \ldots t_n$, $A(t_1), \ldots A(t_n)$ be as in (i). Then given

$$(t_i \neq t_j) \And (\And A(t_i))$$

$(Mv)A(v)$ may be entered as a further line, depending on the same assumptions as $(t_i \neq t_j) \And (\And A(t_i))$.

The idea behind (i) is that if nearly everything has some property A, then in any set of n *distinct* objects, at least one has property A. $t_1, \ldots t_n$ are a set of n terms. If the antecedent of $(t_i \neq t_j) \to \bigvee A(t_i)$ is true, $t_1, \ldots t_n$ all denote distinct objects. But then if nearly everything has A, at least one of the n distinct objects denoted by $t_1, \ldots t_n$ has A. So the disjunction $\bigvee A(t_i)$ is true. And so the whole conditional is true. If on the other hand $t_1, \ldots t_n$ do not all denote distinct objects, $(t_i \neq t_j)$ is false, and the conditional is hence again true.

The idea behind (ii) is that if $t_1, \ldots t_n$ all denote distinct objects, and every one of these objects has some property A, then there are many things which have A.

(iii) Penuniversal introduction (NI)

Let $A(t)$ be a wff containing a term t, let v be a variable that does not occur in $A(t)$, and let $t_1, t_2, \ldots t_n$ be a set of n distinct terms (n strictly greater than 1). Let $A(t_1), A(t_2), \ldots A(t_n)$ be

c

the results of replacing all and only occurrences of t in $A(t)$ by $t_1, t_2, \ldots t_n$ respectively. Let $A(v)$ be the propositional function in v that results from replacing all and only occurrences of t in $A(t)$ by v. Then given

$$(t_i \neq t_j) \to \bigvee A(t_i)$$

$(Nv)A(v)$ may be entered as a further line, depending upon the same assumptions as $(t_i \neq t_j) \to \bigvee A(t_i)$, provided that none of $t_1, \ldots t_n$ occurs in any assumption upon which $(t_i \neq t_j) \to \bigvee A(t_i)$ depends.

NI may look a bit complicated. It is in fact very similar to *UI*. The idea is this. Suppose we are given a proof of $(t_i \neq t_j) \to \bigvee A(t_i)$ from certain assumptions. Any interpretation that makes the assumptions true makes $(t_i \neq t_j) \to \bigvee A(t_i)$ come out true (this follows from the proof, given later, that the rules are sound). Such an interpretation assigns particular objects to $t_1, t_2, \ldots t_n$. But if none of $t_1, t_2, \ldots t_n$ occurs among the assumptions, it makes no difference to the truth of the assumptions *whatever* objects are assigned to $t_1, t_2, \ldots t_n$, provided the rest of the interpretation remains the same. But since $(t_i \neq t_j) \to \bigvee A(t_i)$ is a consequence of the assumptions, it makes no difference to *its* truth what objects are assigned to $t_1, t_2, \ldots t_n$, provided the rest of the interpretation remains the same. So if the assumptions are true, in *any* set of n distinct objects at least one has property A. But then nearly everything has A.

Certain crucial propositions in this heuristic account are as yet unproved, but the explanation should bring out the similarity to *UI*. The last rule for the special quantifiers for the logic of plurality is parallel to *EE* in the way that *NI* is parallel to *UI*. Thus the analogy between the universal and penuniversal quantifier, and that between the existential and plural quantifier, is quite extensively exploited.

(iv) Plural elimination (ME)
Let $A(t)$, v, $A(t_1)$, $A(t_2)$, $\ldots A(t_n)$ and $A(v)$ be as in *(iii)*. Then given a proof of $(Mv)A(v)$ from $A_1, \ldots A_n$ as assumptions, and a proof of C from $(t_i \neq t_j) \mathbin{\&} (\mathbin{\&} A(t_i))$, $B_1, \ldots B_m$, C may be entered as a further line, depending on $A_1, \ldots A_n$, $B_1, \ldots B_m$, provided that none of $t_1, t_2, \ldots t_n$ occurs either in C or in any of $B_1, \ldots B_m$.

(5) THE RULES FOR IDENTITY

(*i*) *Identity introduction* ($= I$)

A wff of the form $t = t$ may be entered as a line at any stage in any derivation, resting on no assumptions.

(*ii*) *Identity elimination* ($= E$)

Let t_1 and t_2 be terms. Let $A(t_1)$ be a wff containing occurrences of t_1, and let $A(t_2)$ be the result of replacing at least one occurrence of t_1 in $A(t_1)$ by t_2. Then, given $A(t_1)$ and $t_1 = t_2$, $A(t_2)$ may be entered as a further line, depending on any assumptions on which either $A(t_i)$ or $t_1 = t_2$ depends.

This completes the statement of the rules. It is almost time to see them in action.

C   Elementary sequents

Proofs of sequents become intolerably long unless some short cuts are taken. For instance, given $(\forall v)A(v)$, where $q_1, \ldots q_p$ are the numbers of the assumptions upon which $(\forall v)A(v)$ rests, $\&A(t_i)$ can always be derived, as follows:

$q_1, \ldots q_p$ (k)	$(\forall v)A(v)$	—
$q_1, \ldots q_p$ $(k+1)$	$A(t_1)$	$(k)UE$
$q_1, \ldots q_p$ $(k+2)$	$A(t_2)$	$(k)UE$
$\ldots\ldots$		
$\ldots\ldots$		
$q_1, \ldots q_p$ $(k+n)$	$A(t_n)$	$(k)UE$
$q_1, \ldots q_p$ $(k+n+1)$	$A(t_1) \& A(t_2)$	$(k+1),(k+2)$ $\& I$
$q_1, \ldots q_p$ $(k+n+2)$	$A(t_1) \& A(t_2) \& A(t_3)$	$(k+n+1),(k+3)$, $\& I$
$\ldots\ldots$		
$\ldots\ldots$		
$q_1, \ldots q_p$ $(k+2n-1)$	$\& A(t_i)$	$(k+2n-2),(k+n)$, $\& I$

In order to shorten the work of writing out such derivations, we shall put the derivation in an abbreviated form, thus

$q_1, \ldots q_p$ (k)	$(\forall v)A(v)$	
$q_1, \ldots q_p$ $(k+1)$	$\& A(t_i)$	(k) UE_n, $\& I_{n-1}$

Similarly, given $\vee A(t_i)$, $(\exists v)A(v)$ can be derived by a sufficient number of successive applications of $\vee E$ and EI. $\vee A(t_i)$ is

$$(\ldots ((A(t_1) \vee A(t_2)) \vee A(t_3)) \vee \ldots \vee A(t_n))$$

$(\exists v)A(v)$ is derived as follows:

$q_1, \ldots q_p$ (k) $(\ldots ((A(t_1) \vee A(t_2)) \vee A(t_3)) \vee \ldots$
$$A(t_n)) \quad —$$

$k+1$ $(k+1)$ $(\ldots (A(t_1) \vee A(t_2)) \vee A(t_3) \vee \ldots$
$$A(t_{n-1}) \quad A$$

$\ldots \ldots \ldots$
$\ldots \ldots \ldots$

$k+n-2$	$(k+n-2)$	$A(t_1) \vee A(t_2)$	A
$k+n-1$	$(k+n-1)$	$A(t_1)$	A
$k+n-1$	$(k+n)$	$(\exists v)A(v)$	$(k+n-1), EI$
$k+n+1$	$(k+n+1)$	$A(t_2)$	A
$k+n+1$	$(k+n+2)$	$(\exists v)A(v)$	$(k+n+1)\ EI$
$k+n-2$	$(k+n+3)$	$(\exists v)A(v)$	$(k+n-2),(k+n-1),(k+n),$ $(k+n+1),(k+n+2) \vee E$
$k+n+4$	$(k+n+4)$	$A(t_3)$	A
$k+n+4$	$(k+n+5)$	$(\exists v)A(v)$	$(k+n+4)\ EI$
$k+n-3$	$(k+n+6)$	$(\exists v)A(v)$	$(k+n-3),(k+n-2),(k+n+3),$ $(k+n+4),(k+n+5), \vee E$

$\ldots \ldots \ldots$
$\ldots \ldots \ldots$

$q_1, \ldots q_p$ $(k+n+3(n-1))(\exists v)A(v)$
$$(k),(k+1),(k+n+3(n-1)-3),$$
$$(k+n+3(n-1)-2),(k+n+3(n-1)-1) \vee E$$

In other words, $(\exists v)A(v)$ is derived from each disjunct, and successively from the disjunctions $A(t_1) \vee A(t_2)$, $(A(t_1) \vee A(t_2)) \vee A(t_3)$, and so on, by successive applications of $\vee E$. To do this we need to make $n-2$ assumptions of successively smaller disjunctions, n further assumptions, one for each of the disjuncts, and then apply EI n times, and $\vee E$ $n-1$ times, making $3n-3$ further lines in all. This whole process is abbreviated to

$q_1, \ldots q_p$ (k) $\bigvee A(t_i)$
$q_1, \ldots q_p$ $(k+1)$ $(\exists v)A(v)$ $(k)EI_n, \vee E_{n-1}$

In fact we shall frequently go further, and omit the numerical subscripts to the citations on the far right.

Other abbreviative conventions are also needed to shorten the work. Suppose we are given $\bigvee A(t_i)$, resting on $q_1, \ldots q_p$, and $\&(A(t_i) \to B(t_i))$, resting on $r_1, \ldots r_s$. Then $\bigvee B(t_i)$ can be derived, depending upon $q_1, \ldots q_p$, $r_1, \ldots r_s$, as follows

$q_1, \ldots q_p$	(j)	$\vee A(t_i)$	
$r_1, \ldots r_s$	(k)	$\&(A(t_i) \to B(t_i))$	
$r_1, \ldots r_s$	$(k+1)$	$A(t_1) \to B(t_1)$	$(k)\ \&\ E$
	....		
	....		
$r_1, \ldots r_s$	$(k+n)$	$A(t_n) \to B(t_n)$	$(k)\ \&\ E$
$k+n+1$	$(k+n+1)$	$(\ldots((A(t_1) \vee A(t_2) \vee$	
		$A(t_3)) \vee \ldots \vee A(t_{n-1})$	A
	....		
	....		
$k+2n-2$	$(k+2n-2)$	$A(t_1) \vee A(t_2)$	A
$k+2n-1$	$(k+2n-1)$	$A(t_1)$	A
$r_1, \ldots r_s, k+2n-1$	$(k+2n)$	$B(t_1)$	$(k+2n-1),(k+1), MPP$
$r_1, \ldots r_s, k+2n-1$	$(k+2n+1)$	$B(t_1) \vee B(t_2)$	$(k+2n)\quad \vee I$
	....		
	....		
$r_1, \ldots r_s, k+2n-1$	$(k+3n-1)$	$\vee B(t_i)$	$(k+3n-2)\quad \vee I$
$k+3n$	$(k+3n)$	$A(t_2)$	A
$r_1, \ldots r_s, k+3n$	$(k+3n+1)$	$B(t_2)$	$(k+3n),(k+2)\quad MPP$
	....		
	....		
$r_1, \ldots r_s, k+3n$	$(k+4n)$	$\vee B(t_i)$	$(k+4n-1)\quad \vee I$
$r_1, \ldots r_s, k+2n-2$	$(k+4n+1)$	$\vee B(t_i)$	$(k+2n-2),(k+2n-1),$
			$(k+3n-1),(k+3n),(k+4n+1)\quad \vee E$

And so on, by successive applications of MPP, $\vee I$, and $\vee E$, until all the assumptions $(k+n+1)$ to $(k+2n-2)$, and all the assumptions of the disjuncts, have been eliminated, giving the conclusion $\vee B(t_i)$ resting on $q_1, \ldots q_p$, $r_1, \ldots r_s$.

In such cases as these, we merely cite on the far right the names of the rules for the connectives necessary for the derivation of the conclusion, thus:

$q_1, \ldots q_p$	(j)	$\vee A(t_i)$	
$r_1, \ldots r_s$	(k)	$\&(A(t_i) \to B(t_i))$	
$q_1, \ldots q_p,\ r_1, \ldots r_s$	$(k+1)$	$\vee B(t_i)$	$(j),(k)\ \&\ E,$
			$MPP, \vee I, \vee E$

When such reasoning has become familiar through examples,

we shall simply cite RC, on the right, standing for 'Rules for Connectives'.

It should be particularly noted that the following generalized De Morgan laws hold

$$\sim(\And\sim A(t_i)) \vdash \bigvee A(t_i), \text{ and vice versa.}$$
$$\sim(\bigvee\sim A(t_i)) \vdash \And A(t_i), \text{ and vice versa.}$$

Similarly

$$\sim(\And A(t_i)) \vdash \bigvee\sim A(t_i), \text{ and vice versa.}$$
$$\sim(\bigvee A(t_i)) \vdash \And\sim A(t_i), \text{ and vice versa.}$$

Such reasoning is common in dealing with n-termed conjunctions and disjunctions. We adopt the short cut of simply passing from one such formula to another, merely citing DM (De Morgan), on the far right.

We now prove a number of sequents, both to illustrate the rules and because some sequents are of interest in their own right. It is all the more important to give several examples, and also to point out certain sequent-expressions that are *not* sequents, because these principles are not available elsewhere. We must therefore ask the reader to bear with the tedium of following through some rather simple proofs.

$(SW1)$ $(Mx)(Fx \And Gx) \vdash (Mx)Fx \And (Mx)Gx$

		Proof	
1	(1)	$(Mx)(Fx \And Gx)$	A
2	(2)	$a_i \neq a_j \And (\And (Fa_i \And Ga_i))$	A
2	(3)	$a_i \neq a_j$	(2) $\And E$
2	(4)	$\And (Fa_i \And Ga_i)$	(2) $\And E$
2	(5)	$\And Fa_i$	(4) $\And E, \And I$
2	(6)	$\And Ga_i$	(4) $\And E, \And I$
2	(7)	$a_i \neq a_j \And (\And Fa_i)$	(3),(5) $\And I$
2	(8)	$a_i \neq a_j \And (\And Ga_i)$	(3),(6) $\And I$
2	(9)	$(Mx)Fx$	(7) MI
2	(10)	$(Mx)Gx$	(8) MI
2	(11)	$(Mx)Fx \And (Mx)Gx$	(9),(10) $\And I$
1	(12)	$(Mx)Fx \And (Mx)Gx$	(1),(2),(11) ME

This proof is very similar to the proof of the corresponding sequent-expression of classical predicate logic.

$(SP1)$ $(\exists x)(Fx \And Gx) \vdash (\exists x)Fx \And (\exists x)Gx$

	Proof	
1	(1) $(\exists x)(Fx \ \& \ Gx)$	A
2	(2) $Fa \ \& \ Ga$	A
2	(3) Fa	(2) E
2	(4) Ga	(2) E
2	(5) $(\exists x)Fx$	(3) EI
2	(6) $(\exists x)Gx$	(4) EI
2	(7) $(\exists x)Fx \ \& \ (\exists x)Gx$	(5),(6) & I
1	(8) $(\exists x)Fx \ \& \ (\exists x)Gx$	(1),(2),(7) EE

The proof of $(SW1)$ is rather longer, because of the need to take account of the non-identity $t_i \neq t_j$.

Just as

(SP^*1) $(\exists x)Fx \ \& \ (\exists x)Gx \vdash (\exists x)(Fx \ \& \ Gx)$

is not a sequent of classical predicate logic, so

(SW^*1) $(Mx)Fx \ \& \ (Mx)Gx \vdash (Mx)(Fx \ \& \ Gx)$

is not a sequent of the system W. (SW^*1) is intuitively invalid in an obvious way.

$(SW2)$ $(Mx)Fx \lor (Mx)Gx \vdash (Mx)(Fx \lor Gx)$

	Proof	
1	(1) $(Mx)Fx \lor (Mx)Gx$	A
2	(2) $(Mx)Fx$	A
3	(3) $a_i \neq a_j \ \& \ (\& \ Fa_i)$	A
3	(4) $a_i \neq a_j$	(3) & E
3	(5) $\& \ Fa_i$	(3) & E
3	(6) $\& \ (Fa_i \lor Ga_i)$	(5) & E, $\lor$ I, & I
3	(7) $a_i \neq a_j \ \& \ (\& \ (Fa_i \lor Ga_i))$	(4),(6) & I
3	(8) $(Mx)(Fx \lor Gx)$	(7) MI
9	(9) $(Mx)Gx$	A
10	(10) $a_i \neq a_j \ \& \ (\& \ Ga_i)$	A
10	(11) $a_i \neq a_j$	(10) & E
10	(12) $\& \ Ga_i$	(10) & E
10	(13) $\& \ (Fa_i \lor Ga_i)$	(12) & E, $\lor$ I, & I
10	(14) $a_i \neq a_j \ \& \ (\& \ (Fa_i \lor Ga_i)$	(11),(13) & I
10	(15) $(Mx)(Fx \lor Gx)$	(14) MI
2	(16) $(Mx)(Fx \lor Gx)$	(2),(3),(8) ME
9	(17) $(Mx)(Fx \lor Gx)$	(9),(10),(15) ME
1	(18) $(Mx)(Fx \lor Gx)$	(1),(2),(16),(9),(17) $\lor$ E

The converse sequent-expression

(SW^*2) $(Mx)(Fx \lor Gx) \vdash (Mx)Fx \lor (Mx)Gx$

is, however, not a sequent. A counter-example to it is this. Suppose that n is the least number of objects that constitute a manifold. Suppose that $k < n$, $m < n$, but $k + m \geq n$, and that there are exactly k things that are F, and exactly m other things with G. Then '$(Mx)(Fx \lor Fx)$' is true, but '$(Mx)Fx \lor (Mx)Gx$' is false. This is a point of contrast between 'many' and 'some', since of course

$(SP2)$ 　$(\exists x)(Fx \lor Gx) \vdash (\exists x)Fx \lor (\exists x)Gx$

is a sequent of classical logic.

$(SW3)$ 　$(Nx)(Fx \& Gx) \vdash (Nx)Fx \& (Nx)Gx$

Proof

1	(1) $(Nx)(Fx \& Gx)$	A
1	(2) $a_i \neq a_j \to \bigvee(Fa_i \& Ga_i)$	(1) NE
3	(3) $a_i \neq a_j$	A
1,3	(4) $\bigvee(Fa_i \& Ga_i)$	(2),(3) MPP
1,3	(5) $\bigvee Fa_i$	(4), $\lor E, \& E, \lor I$
1	(6) $a_i \neq a_j \to \bigvee Fa_i$	(3),(5) CP
1	(7) $(Nx)Fx$	(6) NI
8	(8) $a_i \neq a_j$	A
1,8	(9) $\bigvee(Fa_i \& Ga_i)$	(2),(8) MPP
1,8	(10) $\bigvee Ga_i$	(9), $\lor E, \& E, \lor I$
1	(11) $a_i \neq a_j \to \bigvee Ga_i$	(8),(10) CP
1	(12) $(Nx)Gx$	(11) NI
1	(13) $(Nx)Fx \& (Nx)Gx$	(7),(12) $\& I$

The converse sequent-expression

(SW^*3) $(Nx)Fx \& (Nx)Gx \vdash (Nx)(Fx \& Gx)$

is not a sequent. If n is the least number that constitutes a manifold, and $k < n$, $m < n$, and all but k things have F, and all but m things have G, then '$(Nx)Fx \& (Nx)Gx$' is true. But if $k + m \geq n$, and the k things that do not have F are none of them the same as any of the m things that do not have G, '$(Nx)(Fx \& Gx)$' is false. This is a point of contrast between 'nearly all' and 'all', since of course

$(SP3)$ 　$(\forall x)Fx \& (\forall x)Gx \vdash (\forall x)(Fx \& Gx)$

is a sequent of classical logic.

$(SW4)$ $(Nx)Fx \vee (Nx)Gx \vdash (Nx)(Fx \vee Gx)$

Proof

1	(1)	$(Nx)Fx \vee (Nx)Gx$	A
2	(2)	$(Nx)Fx$	A
2	(3)	$a_i \neq a_j \rightarrow \bigvee Fa_i$	(2) NE
4	(4)	$a_i \neq a_j$	A
2,4	(5)	$\bigvee Fa_i$	(3),(4) MPP
2,4	(6)	$\bigvee(Fa_i \vee Ga_i)$	(5) $\vee I$, $\vee E$
2	(7)	$a_i \neq a_j \rightarrow \bigvee(Fa_i \vee Ga_i)$	(4),(6) CP
2	(8)	$(Nx)(Fx \vee Gx)$	(7) NI
9	(9)	$(Nx)Gx$	A
9	(10)	$a_i \neq a_j \rightarrow \bigvee Ga_i$	(9) NE
11	(11)	$a_i \neq a_j$	A
9,11	(12)	$\bigvee Ga_i$	(10),(11) MPP
9,11	(13)	$\bigvee(Fa_i \vee Ga_i)$	(12) $\vee I$, $\vee E$
9	(14)	$a_i \neq a_j \rightarrow \bigvee(Fa_i \vee Ga_i)$	(11),(13) CP
9	(15)	$(Nx)(Fx \vee Gx)$	(14) NI
1	(16)	$(Nx)(Fx \vee Gx)$	(1),(2),(8),(9),(15), $\vee E$

The converse sequent-expression

(SW^*4) $(Nx)(Fx \vee Gx) \vdash (Nx)Fx \vee (Nx)Gx$

is not a sequent.

Now consider the sequent-expression corresponding to proposition (29) of Chapter 1, Section **C**.

(SW^*5) $(Nx)(Fx \rightarrow Gx), (Nx)Fx \vdash (Nx)Gx$.

It is instructive to give two counter-examples to illustrate the invalidity of this expression.

(1) Let n be the smallest number of elements that constitute a manifold, and consider a domain with exactly n elements. Suppose that none of these elements have G, but that some elements have F, and some do not have F. Then both '$(Nx)Fx$' and '$(Nx)\sim Fx$' are true, since any manifold exhausts the entire domain, and in that manifold there is at least one element with F, and at least one without F. From '$(Nx)\sim Fx$' there follows '$(Nx)(Fx \rightarrow Gx)$', so the latter is also true. Thus both expressions on left of '$\vdash$' in (SW^*4) are true. But '$(Nx)Gx$' is false. This

counter-example is interesting in that it illustrates the consistency of '$(Nx)Fx$' with '$(Nx)\sim Fx$'. The consistency of this pair is comparable to that of '$(\forall x)Fx$' with '$(\forall x)\sim Fx$' in an empty domain. For '$(Nx)Fx$' and '$(Nx)\sim Fx$' cannot both be true in a domain that is sufficiently large in relation to the least number of elements that constitute a manifold. In fact, if '$(Nx)Fx$' is true, at most $n-1$ elements do not have F, and if '$(Nx)\sim Fx$' is true, at most $n-1$ elements have F. Any element either has F or does not have F, so if both '$(Nx)Fx$' and '$(Nx)\sim Fx$' are true, there are at most $2n-2$ elements in the domain.

(2) Suppose all but $n-1$ elements have F, so that '$(Nx)Fx$' is true. Suppose that all but exactly n elements have G, so that '$(Nx)Gx$' is false. Suppose that the $n-1$ elements that do not have F also do not have G. Then there is only one element that has F and does not have G. So '$(Nx)(Fx \rightarrow Gx)$' is true. Then the expressions to the left of '$\vdash$' in $(SW*4)$ are true, but the expression to the right is false. $(SW*4)$ is therefore not a valid principle of inference.

The situation may be made clear by a diagram.

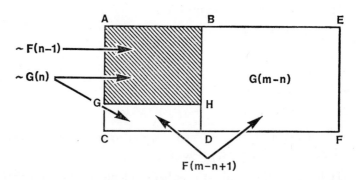

The rectangle $AECF$ represents the domain of discourse, containing m elements. $ABCD$ represents the n elements that are not G. The shaded rectangle $ABGH$ represents the $n-1$ elements which are not F.

A counter-example in terms of real propositions might be this. Suppose the domain is that of philosophers. Nearly all who know some logic have read Frege. Nearly all know some logic. But it is false that nearly all have read Frege.

We now prove a principle used in counter-example (1).

$(SW5)$ $(Nx)\sim Fx \vdash (Nx)(Fx \to Gx)$

       Proof

1	(1) $(Nx)\sim Fx$	A
1	(2) $a_i \neq a_j \to \vee \sim Fa_i$	(1) NE
3	(3) $a_i \neq a_j$	A
1,3	(4) $\vee \sim Fa_i$	(2),(3) MPP
1,3	(5) $\vee (Fa_i \to Ga_i)$	(4) RC
1	(6) $a_i \neq a_j \to \vee (Fa_i \to Ga_i)$	(3),(5) CP
1	(7) $(Nx)(Fx \to Gx)$	(6) NI

Here at line (5) we begin the practice of citing simply RC on the right, where it is clear that the line follows by the rules for the connectives.

The next task is to prove the laws of interchange of plural and penuniversal quantifiers.

$(SW6)$ $(Nx)Fx \vdash \sim(Mx)\sim Fx$

       Proof

1	(1) $(Nx)Fx$	A
2	(2) $(Mx)\sim Fx$	A
3	(3) $a_i \neq a_j \,\&\, (\&\sim Fa_i)$	A
1	(4) $a_i \neq a_j \to \vee Fa_i$	(1) NE
3	(5) $a_i \neq a_j$	(3) $\&\,E$
1,3	(6) $\vee Fa_i$	(4),(5) MPP
1,3	(7) $\sim(\&\sim Fa_i)$	(6) DM
3	(8) $\&\sim Fa_i$	(3) $\&\,E$
1,3	(9) $(\&\sim Fa_i) \,\&\, \sim(\&\sim Fa_i)$	(7),(8) $\&\,I$
3	(10) $\sim(Nx)Fx$	(1),(9) RAA
2	(11) $\sim(Nx)Fx$	(2),(3),(10), ME
1,2	(12) $(Nx)Fx \,\&\, \sim(Nx)Fx$	(1),(11) $\&\,I$
1	(13) $\sim(Mx)\sim Fx$	(2),(12) RAA

$(SW7)$ $\sim(Mx)\sim Fx \vdash (Nx)Fx$

       Proof

1	(1) $\sim(Mx)\sim Fx$	A
2	(2) $a_i \neq a_j \,\&\, (\&\sim Fa_i)$	A
2	(3) $(Mx)\sim Fx$	(2) MI
1,2	(4) $(Mx)\sim Fx \,\&\, \sim(Mx)\sim Fx$	(1),(3) $\&\,I$

1	(5) $\sim(a_i \neq a_j \ \& \ (\& \sim Fa_i))$	(2),(4) RAA
1	(6) $a_i \neq a_j \rightarrow \sim(\& \sim Fa_i)$	(5) RC
7	(7) $a_i \neq a_j$	A
1,7	(8) $\sim(\& \sim Fa_i)$	(6),(7) MPP
1,7	(9) $\vee Fa_i$	(8) DM
1	(10) $a_i \neq a_j \rightarrow \vee Fa_i$	(7),(9) CP
1	(11) $(Nx)Fx$	(10) NI

$(SW8)$　　　$(Mx)Fx \vdash \sim(Nx)\sim Fx$

Proof

1	(1) $(Mx)Fx$	A
2	(2) $a_i \neq a_j \ \& \ (\& \ Fa_i)$	A
3	(3) $(Nx)\sim Fx$	A
3	(4) $a_i \neq a_j \rightarrow \vee \sim Fa_i$	(3) NE
2	(5) $a_i \neq a_j$	(2) $\& E$
2,3	(6) $\vee \sim Fa_i$	(4),(5) MPP
2	(7) $\& \ Fa_i$	(2) $\& E$
2	(8) $\sim(\vee \sim Fa_i)$	(7) DM
2,3	(9) $\vee \sim Fa_i \ \& \ \sim(\vee \sim Fa_i)$	(6),(8) $\& I$
2	(10) $\sim(Nx)\sim Fx$	(3),(9) RAA
1	(11) $\sim(Nx)\sim Fx$	(1),(2),(10) ME

$(SW9)$　　　$\sim(Nx)\sim Fx \vdash (Mx)Fx$

Proof

1	(1) $\sim(Nx)\sim Fx$	A
2	(2) $\sim(Mx)Fx$	A
3	(3) $a_i \neq a_j \ \& \ (\& \ Fa_i)$	A
3	(4) $(Mx)Fx$	(3) MI
2,3	(5) $(Mx)Fx \ \& \ \sim(Mx)Fx$	(2),(4) $\& I$
2	(6) $\sim(a_i \neq a_j \ \& \ (\& \ Fa_i))$	(3),(5) RAA
2	(7) $a_i \neq a_j \rightarrow \vee \sim Fa_i$	(6) RC, DM
2	(8) $(Nx)\sim Fx$	(7) NI
1,2	(9) $(Nx)\sim Fx \ \& \ \sim(Nx)\sim Fx$	(1),(8) $\& I$
1	(10) $\sim\sim(Mx)Fx$	(2),(9) RAA
1	(11) $(Mx)Fx$	(10) DN

Proofs of the following sequents are left as exercises:

$(SW10)$　$(Mx)\sim Fx \vdash \sim(Nx)Fx$
$(SW11)$　$\sim(Nx)Fx \vdash (Mx)\sim Fx$
$(SW12)$　$\sim(Mx)Fx \vdash (Nx)\sim Fx$

$(SW13)$ $(Nx)\sim Fx \vdash \sim(Mx)Fx.$

$(SW14)$ $(Nx)(Fx \to P) \vdash (Mx)Fx \to P$

Proof

1	(1) $(Nx)(Fx \to P)$	A
2	(2) $(Mx)Fx$	A
3	(3) $a_i \neq a_j \;\&\; (\&\; Fa_i)$	A
1	(4) $a_i \neq a_j \to \bigvee(Fa_i \to P)$	(1) NE
3	(5) $a_i \neq a_j$	(3) $\& E$
1,3	(6) $\bigvee(Fa_i \to P)$	(4),(5) MPP
3	(7) $\& \; Fa_i$	(3) $\& E$
1,3	(8) P	(6),(7) RC
1,2	(9) P	(2),(3),(8) ME
1	(10) $(Mx)Fx \to P$	(2),(9) CP

$(SW15)$ $(Mx)Fx \to P \vdash (Nx)(Fx \to P)$

Proof

1	(1) $(Mx)Fx \to P$	A
2	(2) $a_i \neq a_j \;\&\; (\&\; Fa_i)$	A
2	(3) $(Mx)Fx$	(2) MI
1,2	(4) P	(1),(2) MPP
1	(5) $((a_i \neq a_j) \;\&\; (\&\; Fa_i)) \to P$	(2),(4) CP
1	(6) $(a_i \neq a_j) \to ((\&\; Fa_i) \to P$	(5) RC
7	(7) $a_i \neq a_j$	A
1,7	(8) $(\&\; Fa_i) \to P$	(6),(7) MPP
1,7	(9) $\bigvee(Fa_i \to P)$	(8) RC
1	(10) $a_i \neq a_j \to \bigvee(Fa_i \to P)$	(7),(9) CP
1	(11) $(Nx)(Fx \to P)$	(10) NI

Proofs of the following sequents are left as exercises:

$(SW16)$ $(Mx)(P \to Fx) \vdash P \to (Mx)Fx$
$(SW17)$ $P \to (Mx)Fx \vdash (Mx)(P \to Fx)$
$(SW18)$ $(Nx)(P \to Fx) \vdash P \to (Nx)Fx$
$(SW19)$ $P \to (Nx)Fx \vdash (Nx)(P \to Fx).$

$(SW20)$ $(Mx)(Fx \to P) \vdash (Nx)Fx \to P$

Proof

1	(1) $(Mx)(Fx \to P)$	A
2	(2) $(Nx)Fx$	A
3	(3) $a_i \neq a_j \;\&\; (\&\; (Fa_i \to P))$	A

2	(4) $a_i \neq a_j \rightarrow \bigvee Fa_i$	(2) NE
3	(5) $a_i \neq a_j$	(3) $\& E$
3	(6) $\& (Fa_i \rightarrow P)$	(3) $\& E$
2,3	(7) $\bigvee Fa_i$	(4),(5) MPP
2,3	(8) P	(6),(7) RC
1,2	(9) P	(1),(3),(8) ME
1	(10) $(Nx)Fx \rightarrow P$	(2),(9) CP

$(SW21)$ $(Nx)Fx \rightarrow P \vdash (Mx)(Fx \rightarrow P)$

Proof of $(SW21)$ is left as an exercise.

In the next section we turn to sequent-expressions involving both classical quantifiers and quantifiers specific to the logic of plurality.

D   Sequents with both classical and plural quantifiers

It might be thought that it should be possible to prove that if everything has some property F, then many things have some property F. That is, it might be thought that the following should be a sequent

$(SW*6)$ $(\forall x)Fx \vdash (Mx)Fx$.

But $(SW*6)$ is not valid in the system W. For in W, we wish to be able to prove only such principles as are valid in every non-empty domain. In a domain with fewer than n members, all of which have F, '$(\forall x)Fx$' is true, but '$(Mx)Fx$' is false. We can derive '$\& Fa_i$' from '$(\forall x)Fx$', but we cannot get to '$(Mx)Fx$' without the extra premiss '$a_i \neq a_j$'. On the other hand, given that extra premiss, the conclusion '$(Mx)Fx$' *is* derivable. So we have

$(SW22)$ $(\forall x)Fx, a_i \neq a_j \vdash (Mx)Fx$

> *Proof*

1	(1) $(\forall x)Fx$	A
2	(2) $a_i \neq a_j$	A
1	(3) $\& Fa_i$	(1) UE_n
1,2	(4) $a_i \neq a_j \& (\& Fa_i)$	(2),(3) $\& I$
1,2	(5) $(Mx)Fx$.	(4) MI

Similarly, any extra premiss which ensures that any domain in

which it is true has many members will do in place of '$a_i \neq a_j$'.
So we have

$(SW23)$ $(\forall x)Fx, (Mx)Gx \vdash (Mx)Fx$

The proof is left as an exercise.

Again, it might be thought that if nearly everything has F,
then something has F, i.e. that the following should be a sequent

$(SW^{*}7)$ $(Nx)Fx \vdash (\exists x)Fx$

But $(SW^{*}7)$ is not a sequent. '$(Nx)Fx$' is satisfied (vacuously)
in any domain with fewer than n elements, whether or not any
of those elements have property F. On the other hand
$(SW24)$ $(Nx)Fx, a_i \neq a_j \vdash (\exists x)Fx$
is a sequent. We leave the proof of that to the reader, and
prove instead

$(SW25)$ $(Nx)Fx, (Mx)Gx \vdash (\exists x)Fx$

Proof

1	(1) $(Nx)Fx$	A
2	(2) $(Mx)Gx$	A
3	(3) $a_i \neq a_j \,\&\, (\& \, Ga_i)$	A
1	(4) $a_i \neq a_j \rightarrow \vee Fa_i$	(1) NE
3	(5) $a_i \neq a_j$	(3) $\& E$
1,3	(6) $\vee Fa_i$	(4),(5) MPP
1,3	(7) $(\exists x)Fx$	(6) $EI, \vee E$
1,2	(8) $(\exists x)Fx$	(2),(3),(7) ME

In a domain with fewer than n elements, any wff of the form
$(Nv)A(v)$ is satisfied vacuously. But certain wffs of this form
can only be satisfied vacuously, for instance '$(Nx)(Fx \,\&\sim Fx)$'.
If such a wff is true, the domain must contain fewer than n
elements. So in that domain, *no* wff of the form $(Mv)A(v)$ is true.
The next sequent makes this point.

$(SW26)$ $(Nx)(Fx \,\&\sim Fx) \vdash \, \sim(Mx)Gx$

Proof

1	(1) $(Nx)(Fx \,\&\sim Fx)$	A
2	(2) $a_i \neq a_j \,\&\, (\& \, Ga_i)$	A
1	(3) $a_i \neq a_j \rightarrow \vee(Fa_i \,\&\sim Fa_i)$	(1) NE
2	(4) $a_i \neq a_j$	(2) $\& E$

1,2	(5) $\vee (Fa_i \ \& \sim Fa_i)$	(3),(4) MPP
1	(6) $\sim (a_i \neq a_j \ \& \ (\& \ Ga_i))$	(2),(5) $\vee E, RAA$
1	(7) $a_i \neq a_j \rightarrow \sim (\& \ Ga_i)$	(6) RC
1	(8) $a_i \neq a_j \rightarrow \vee \sim Ga_i$	(7) RC, DM
1	(9) $(Nx)\sim Gx$	(8) NI
1	(10) $\sim (Mx)Gx$	(9) by $(SW13)$

The converse principle to $(SW26)$ is rather obvious. It is

$(SW27)$ $(Mx)Gx \vdash (Mx)(Fx \vee \sim Fx)$

Proof is left to the reader.

$(SW28)$ $(Nx)(Fx \rightarrow Gx), (Mx)Fx \vdash (\exists x)Gx$

Proof

1	(1) $(Nx)(Fx \rightarrow Gx)$	A
2	(2) $(Mx)Fx$	A
3	(3) $a_i \neq a_j \ \& \ (\& \ Fa_i)$	A
1	(4) $a_i \neq a_j \rightarrow \vee (Fa_i \rightarrow Ga_i)$	(1) NE
3	(5) $a_i \neq a_j$	(3) $\& \ E$
1,3	(6) $\vee (Fa_i \rightarrow Ga_i)$	(4),(5) MPP
3	(7) $\& \ Fa_i$	(3) $\& \ E$
1,3	(8) $\vee Ga_i$	(6),(7) RC
1,3	(9) $(\exists x)Gx$	(8) $EI, \vee E$
1,2	(10) $(\exists x)Gx$	(2),(3),(9) ME

$(SW29)$ $(Nx)(Fx \rightarrow Gx), (Mx)\sim Gx \vdash (\exists x)\sim Fx$

Proof

1	(1) $(Nx)(Fx \rightarrow Gx)$	A
2	(2) $(Mx)\sim Gx$	A
3	(3) $a_i \neq a_j \ \& \ (\& \sim Ga_i)$	A
1	(4) $a_i \neq a_j \rightarrow \vee (Fa_i \rightarrow Ga_i)$	(1) NE
3	(5) $a_i \neq a_j$	(3) $\& \ E$
1,3	(6) $\vee (Fa_i \rightarrow Ga_i)$	(4),(5) MPP
3	(7) $\& \sim Ga_i$	(3) $\& \ E$
1,3	(8) $\vee \sim Fa_i$	(6),(7) RC
1,3	(9) $(\exists x)\sim Fx$	(8) $EI, \vee E$
1,2	(10) $(\exists x)\sim Fx$	(2),(3),(9) ME

$(SW30)$ $(\forall x)(Fx \rightarrow Gx), (Mx)Fx \vdash (Mx)Gx$

Proof

1	(1) $(\forall x)(Fx \rightarrow Gx)$	A

2	(2) $(Mx)Fx$	A
3	(3) $a_i \neq a_j \ \& \ (\& \ Fa_i)$	A
1	(4) $\& \ (Fa_i \rightarrow Ga_i)$	(1) $UE, \ \& \ I$
3	(5) $\& \ Fa_i$	(3) $\& \ E$
1,3	(6) $\& \ Ga_i$	(4),(5) RC
3	(7) $a_i \neq a_j$	(3) $\& \ E$
1,3	(8) $a_i \neq a_j \ \& \ (\& \ Ga_i)$	(6),(7) $\& \ I$
1,3	(9) $(Mx)Gx$	(8) MI
1,2	(10) $(Mx)Gx$	(2),(3),(9) ME

$(SW29) \quad (Nx)(Fx \rightarrow Gx), (\forall x)Fx \vdash (Nx)Gx$

Proof

1	(1) $(Nx)(Fx \rightarrow Gx)$	A
2	(2) $(\forall x)Fx$	A
1	(3) $a_i \neq a_j \rightarrow \bigvee (Fa_i \rightarrow Ga_i)$	(1) NE
4	(4) $a_i \neq a_j$	A
1,4	(5) $\bigvee (Fa_i \rightarrow Ga_i)$	(3),(4) MPP
2	(6) $\& \ Fa_i$	(2) $UE, \ \& \ I$
1,2,4	(7) $\bigvee Ga_i$	(5),(6) RC
1,2	(8) $a_i \neq a_j \rightarrow \bigvee Ga_i$	(4),(7) CP
1,2	(9) $(Nx)Gx$	(8) NI

$(SW30) \quad (\forall x)(Fx \rightarrow Gx), (Nx)Fx \vdash (Nx)Gx$

Proof of $(SW30)$ is left to the reader, as are proofs of the following sequents:

$(SW31) \quad (\forall x)Fx \vdash (Nx)Fx$

$(SW32) \quad (Mx)\sim Fx \vdash \sim (\forall x)Fx$

$(SW33) \quad (Mx)(Fx \vee Gx), \sim (\exists x)Fx \vdash (Mx)Gx$

$(SW34) \quad (\forall x)(Fx \ \& \ Gx), (Mx)Fx \vdash (Mx)Gx.$

E Arguments with propositional functions in more than one variable

We begin with an old friend from Chapter 1.

$(SW*8) \quad (Mx)(My)Fxy \vdash (My)(Mx)Fxy.$

$(SW*8)$ is not a sequent, but the following *is*

$(SW35) \quad (Mx)(\forall y)Fxy \vdash (My)(Mx)Fxy$

Proof

1	(1) $(Mx)(\forall y)Fxy$	A
2	(2) $a_i \neq a_j \ \& \ (\& \ (\forall y)Fa_iy)$	A

D

2	(3)	$\& (\forall y) Fa_i y$	(2) $\& E$
2	(4)	$\underset{1 \le j \le n}{\&} \quad (\underset{1 \le i \le n}{\&} \ Fa_i a_j)$	(3), $\& E$, UE, $\& I$
2	(5)	$a_i \ne a_j$	(2) $\& E$
2	(6)	$\underset{1 \le i \le n}{\&} \quad Fa_i a_1$	(4) $\& E$
2	(7)	$a_i \ne a_j \ \& \ (\underset{1 \le i \le n}{\&} \ Fa_i a_1)$	(5),(6) $\& I$
2	(8)	$(Mx) Fxa_1$	
2	(9)	$\underset{1 \le i \le n}{\&} \quad Fa_i a_2$	(4) $\& E$
2	(10)	$a_i \ne a_j \ \& \ (\& \ Fa_i a_2)$	(5),(9) $\& I$
2	(11)	$(Mx) Fxa_2$	(10) MI

. . . .
. . . .

2	$(5+3n)$	$(Mx) Fxa_n$	$(5+3n-1)$ MI
2	$(5+3n+1)$	$\& (Mx) Fxa_i$	(8),(11) $\ldots (5+3n)$ $\& I$
2	$(5+3n+2)$	$a_i \ne a_j \& (\&(Mx) Fxa_i)$	(2),$(5+3n+1)\& I$
2	$(5+3n+3)$	$(My)(Mx) Fxy$	$(5+3n+2)$ MI
1	$(5+3n+4)$	$(My)(Mx) Fxy$	(1),(2),$(5+3n+3)$ ME

The expression

$(SW*9) \quad (\forall x)(\forall y) Fxy \vdash (My)(Mx) Fxy$

is not a sequent, but we do have

$(SW36) \quad (\forall x)(\forall y) Fxy, (Mx) Gx \vdash (My)(Mx) Fxy$

Proof

1	(1)	$(\forall x)(\forall y) Fxy$	A
2	(2)	$(Mx) Gx$	A
3	(3)	$a_i \ne a_j \ \& \ (\& \ Ga_i)$	A
1	(4)	$\& (\forall y) Fa_i y$	(1) UE, $\& I$
1	(5)	$\underset{1 \le j \le n}{\&} \quad (\underset{1 \le i \le n}{\&} \ Fa_i a_j)$	(4) UE, $\& E$, $\& I$
1	(6)	$\underset{1 \le i \le n}{\&} \quad (\underset{1 \le j \le n}{\&} \ Fa_i a_j)$	(5) RC
3	(7)	$a_i \ne a_j$	(3) $\& E$
1,3	(8)	$a_i \ne a_j \ \& \ (\underset{1 \le i \le n}{\&} \ (\underset{1 \le j \le n}{\&} \ Fa_i a_j))$	(6),(7) $\& I$
1,3	(9)	$(Mx) \underset{1 \le j \le n}{\&} \quad Fxa_j$	(8) MI

1,3	(10)	$\&\,(Mx)Fxa_j$	(9), by $(SW1)$
1,3	(11)	$a_i \neq a_j \;\&\; (\&\,(Mx)Fxa_i)$	(7),(10) $\&\,I$
1,3	(12)	$(My)(Mx)Fxy$	(11) MI
1,2	(13)	$(My)(Mx)Fxy$	(2),(3),(12) ME

(5) is an n-termed conjunction, each of whose conjuncts is an n-termed conjunction. (6) results from (5) by rearranging the conjuncts.

$(SW37) \quad (\exists y)(Mx)Fxy \vdash (Mx)(\exists y)Fxy$

$\qquad$ *Proof*

1	(1)	$(\exists y)(Mx)Fxy$	A
2	(2)	$(Mx)Fxb$	A
3	(3)	$a_i \neq a_j \;\&\; (\&\, Fa_ib)$	A
3	(4)	$\&\, Fa_ib$	(3) $\&\,E$
3	(5)	$\&\,(\exists y)Fa_iy$	(4) $\&\,E$, EI, $\&\,I$
3	(6)	$a_i \neq a_j$	(3) $\&\,E$
3	(7)	$a_i \neq a_j \;\&\; (\&\,(\exists y)Fa_iy)$	(5),(6) $\&\,I$
3	(8)	$(Mx)(\exists y)Fxy$	(7) MI
2	(9)	$(Mx)(\exists y)Fxy$	(2),(3),(8) ME
1	(10)	$(Mx)(\exists y)Fxy$	(1),(2),(9) EE

$(SW38) \quad (Mx)(\exists y)(Fx \vee Gy) \vdash (Mx)Fx \vee (\exists y)Gy$

Proof is left to the reader.

It is also possible to prove some more general laws of quantifier-interchange, such as

$(SW39) \quad \sim(\forall x)(Ny)Fxy \vdash (\exists x)(My)\sim Fxy$
$(SW40) \quad \sim(\forall x)(My)Fxy \vdash (\exists x)(Ny)\sim Fxy$
$(SW41) \quad \sim(Mx)(\exists y)Fxy \vdash (Nx)(\forall y)\sim Fxy$

and the converses of these. Proofs are again left to the reader.

The sequents we have given should suffice to illustrate the sort of thing that can be proved in the system W, and to give an idea of how to work with the rules.

F   Reduction in the number of rules

It is well known that the set of rules given in this chapter for classical predicate logic contains more rules than are strictly necessary. The rules for the connectives can be reduced to a smaller set in various ways, which we shall not go into, and the rules for the classical quantifiers are also more than are strictly

needed. In particular, if we *define* the existential quantifier in
terms of the universal quantifier, thus

$$\text{Def } \exists\colon (\exists v)A(v) \;=\; \sim(\forall v)\sim A(v)$$

the rules EI and EE can be derived from UE and UI together
with the rules for the connectives. That is to say, any sequent
provable with the help of EI or EE is also provable using only
UE, UI, Def $\exists$, and the rules for the connectives (RC). Such a
reduction is shown clearly in Benson Mates's *Elementary Logic*,
and the proof is not repeated here.

The rules for the quantifiers specific to the logic of plurality
are also more than are needed. If the plural quantifier is defined
in terms of the penuniversal quantifier, thus

$$\text{Def } M\colon (Mv)A(v) \;=\; \sim(Nv)\sim A(v)$$

the rules MI and ME can be derived from NE and NI together
with RC. If both Def $\exists$ and Def M are adopted, the set of rules
can be reduced to RC together with UE, UI, NE and NI.
Here we show that ME and MI can be proved as derived rules,
and assume the reduction of EI and EE to UE and UI. This
reduction in the rules simplifies the discussion of their soundness,
for if any sequent provable with the help of ME and MI is
already provable using only NE and NI together with the rules
of classical logic, it follows that if NE and NI are sound, so are
ME and MI.

MI

We need to show that, given $t_i \neq t_j$ & (& $(A(t_i))$ resting on
certain assumptions $n_1, \ldots n_p$, we can derive $(Mv)A(v)$ on the
same assumptions, using at most RC, NE, NI, UE and UI.
The following derivation-schema suffices to show how this can
be done.

$n_1, \ldots n_p$	(i)	$t_i \neq t_j$ & (& $A(t_i)$)	
j	(j)	$(Nv)\sim A(v)$	A
j	$(j+1)$	$t_i \neq t_j \to \vee \sim A(t_i)$	(j) NE
$n_1, \ldots n_p$	$(j+2)$	$t_i \neq t_j$	(i) & E
$n_1, \ldots n_p, j$	$(j+3)$	$\vee \sim A(t_i)$	$(j+1)(j+2)$ MPP
$n_1, \ldots n_p$	$(j+4)$	& $A(t_i)$	(i) & E
$n_1, \ldots n_p$	$(j+5)$	$\sim(\vee \sim A(t_i)$	$(j+4)$ DM

$n_1, \ldots n_p, j$ $(j+6) \lor \sim A(t_i) \ \& \ \sim(\lor \sim A(t_i))$ $(j+3)(j+5) \ I$

$n_1, \ldots n_p$ $(j+7) \ \sim(Nv)\sim A(v)$ $(j),(j+6) \ RAA$

$n_1, \ldots n_p$ $(j+8) \ (Mv)A(v)$ $(j+7) \ \text{Def } M$

ME

We need to show that, given $(Mv)A(v)$, resting on $n_1, \ldots n_p$, and a proof of C from $t_i \neq t_j \ \& \ (\& \ A(t_i))$, and $m_1, \ldots m_q$ as assumptions, where no t_i occurs in C or in any of $m_1, \ldots m_q$, a proof can be found for C from $(Mv)A(v)$ and $m_1, \ldots m_q$, which uses at most RC, NE, NI, UE and UI. The following derivation-schema suffices.

$n_1, \ldots n_p$ (i) $(Mv)A(v)$

j (j) $t_i \neq t_j \ \& \ (\& \ A(t_i))$ A

. . . .

. . . .

$m_1, \ldots m_q, j$ (k) C

$k+1$ $(k+1) \ \sim C$ A

$m_1, \ldots m_q, j, k+1$ $(k+2) \ C \ \& \ \sim C$

$m_1, \ldots m_q, k+1$ $(k+3) \ \sim(t_i \neq t_j \ \& \ (\& \ A(t_i)))$ $(j)(k+2) \ RAA$

$m_1, \ldots m_q, k+1$ $(k+4) \ t_i \neq t_j \to \lor \sim A(t_i)$ $(k+3) \ RC$

$m_1, \ldots m_q, k+1$ $(k+5) \ (Nv)\sim A(v)$ $(k+4) \ NI$

$n_1, \ldots n_p$ $(k+6) \ \sim(Nv)\sim A(v)$ $(i) \ \text{Def } M$

$n_1, \ldots n_p, m_1, \ldots m_q, k+1$

   $(k+7) \ (Nv)\sim A(v) \ \& \ \sim(Nv)\sim A(v)$

         $(k+5)(k+6) \ \& \ I$

$n_1, \ldots n_p, m_1, \ldots m_q$

   $(k+8) \ \sim\sim C$ $(k+1)(k+7) \ RAA$

$n_1, \ldots n_p, m_1, \ldots m_q$

   $(k+9) \ C$ $(k+8) \ DN$

Notice that since by hypothesis no t_i occurs in any of $m_1, \ldots m_q$, or in $\sim C$ the application of NI to $t_i \neq t_j \to \lor \sim A(t_i)$ at line $(k+4)$ is permitted.

G   Interpretation, validity, soundness of the rules

We begin by making some remarks on the general character of the logic.

Strictly speaking, we are dealing with more than a single system in treating the system W. In the formulation of the rules ME, MI, NE, NI, there is mention of the number n, which must be greater than one. Now this number must be the same for all four rules, and it is a number that is *fixed* for the whole system. A different system is obtained for every choice of n; there is in fact a denumerable sequence of systems W_2, $W_3, \ldots W_m, \ldots$, depending on the choice of n. For instance, a sequent of W_2 is

(SW_21) $\quad (Nx)Fx \vdash a_1 \neq a_2 \rightarrow (Fa_1 \lor Fa_2)$

but this is *not* a sequent of W_3, or of any W_m $(m > 2)$. Similarly

(SW_22) $\quad a_1 \neq a_2 \ \& \ (Fa_1 \ \& \ Fa_2) \vdash (Mx)Fx$

is a sequent of W_2, but of no system with a larger subscript. In W_2, any set with at least two members constitutes a manifold, but in W_3, a set must have at least *three* members to constitute a manifold, so neither (SW_21) nor (SW_22) is valid in W_3. There are also some sequents containing no proper names (purely general sequents), which hold in some systems but not in others. For instance, in W_2, but in no other system, we have

(SW_23) $\quad (Nx)Fx \vdash (\exists x)(\exists y)(x \neq y \rightarrow (Fx \lor Fy)$

and

(SW_24) $\quad (\exists x)(\exists y)(x \neq y \ \& \ (Fx \ \& \ Fy)) \vdash (Mx)Fx$

but these are sequents of no other system W_m.

It has been possible to speak loosely of *the* system W in previous sections because there is a large number of sequents that are common to *all* the systems, and it is these that have engaged our interest as being most characteristic of the logic of plurality. When treating these sequents, it is possible to leave the number n unspecified, and this is what we have done. The reader can verify that the proofs given go through irrespective of *which* W_m we are working in.

There is a close relationship between the plural quantifier and the *numerically definite* quantifiers '$(\exists x)$', read 'there are exactly n objects such that ...'. The numerically definite quantifiers are recursively defined as follows:

$$(\exists x)Fx = \ \sim(\exists x)Fx$$
$$\quad 0$$

$$(\exists x)Fx = (\exists x)(Fx \ \& \ (\ \exists \ y)(Fy \ \& \ y \neq x)).$$
$$\quad n \qquad\qquad\qquad\qquad\quad n-1$$

Now it is also possible recursively to define a sequence of numerical quantifiers '$(\exists_n x)$', to be read 'there are *at least n* objects such that . . . '. The definition is:

$$(\exists_1 x)Fx = (\exists x)Fx$$
$$(\exists_n x)Fx = (\exists x)(Fx \,\&\, (\exists_{n-1}y)(Fy \,\&\, y \neq x)).$$

Now in the system W_n, $(Mv)A(v)$ has exactly the same meaning as $(\exists_n v)A(v)$. And $(Nv)A(v)$ has the same meaning as $\sim(\exists_n v)\sim A(v)$. Consequently, the system W_n can be looked upon as the logic of the numerical quantifier '$(\exists_n x)$'. And the characteristic principles of the logic of plurality are simply the principles that hold for '$(\exists_n x)$', irrespective of the value of n. Now '$(\exists_1 x)$' is simply the ordinary existential quantifier. If we were to allow the case $n = 1$ in the logic of plurality, it could be seen that classical predicate logic is but a limiting case of the logic of plurality. For if $n = 1$, $(Mv)A(v)$ collapses to the existential quantifier, and $(Nv)A(v)$ collapses to the universal quantifier. In fact, the only reason for excluding the case $n = 1$ was the intuitive one, that nobody would ever agree that many things have F if only one thing has F. But admission of this case does not alter the formal structure, since the principles that hold in *every* $W_n(n > 1)$ also hold where $n = 1$, i.e. in W_1.

Now in everyday argument and discussion, we speak of 'many', 'few' and 'nearly all' where either we do not know the exact number of things with certain properties, or it does not matter to us exactly how many there are with certain properties. But we can still argue validly from premisses containing 'many', 'few' or 'nearly all'. This is because there are principles – the characteristic principles of the logic of plurality – that hold irrespective of the exact number of things having the properties in question. So although premisses of this kind may be considered vague in comparison with the corresponding numerical propositions, argument from them can be perfectly precise, since it leads to conclusions that are vague in a corresponding way. Here is a classic example of how concepts, which are by certain standards vague, may yet have a perfectly exact logic.

The logic of plurality, therefore, formalizes an area of informal argument, and is also a more general logic of which classical predicate logic is a special case.

We now go on to discuss the *interpretation* of the wffs.

An *interpretation* is a non-empty set D and a function that assigns to every term an element of D, to every predicate letter of degree n a set of n-tuples belonging to D, and to every propositional variable one of the two *truth-values*, T or F.

For each system W_n, the concept of *satisfaction* can be recursively defined for each interpretation I. Before giving this definition, an auxiliary notion is convenient. Where t is a term, if I' is like I except at most in what it assigns to t, then I' is described as a *t-variant* of I. Clearly all t-variants of I have the same set (or *domain*) D, and I is a t-variant of itself.

Now let A be a wff of W_n, v be a variable, and t the first term not occurring in A. (We suppose the terms are listed in the order $a, b, c, d, a_1, b_1, \ldots$). And let I be an interpretation.

Then,

(1) If A is a propositional variable, I satisfies A if and only if (iff) I assigns the truth-value T to A.

(2) If A is an atomic sentence $Kt_1, \ldots t_n$, then I satisfies A iff the objects I assigns to $t_1, \ldots t_n$, when taken in that order, belong to the set of n-tuples I assigns to K.

(3) If A is an atomic sentence $t_1 = t_2$, then I satisfies A iff the object I assigns to t_1 is the same as the object I assigns to t_2.

(4) If A is $\sim B$, then I satisfies A iff I does not satisfy B.

(5) If A is $(B \& C)$, then I satisfies A iff I satisfies both B and C.

(6) If A is $(B \vee C)$, then I satisfies A iff I satisfies B, or satisfies C (or both).

(7) If A is $(B \to C)$, then I satisfies A iff I does not satisfy B, or satisfies C (or both).

(8) If A is $(\forall v)B(v)$, then I satisfies A iff every t-variant of I satisfies $B(t)$.

(9) If A is $(\exists v)B(v)$, then I satisfies A iff some t-variant of I satisfies $B(t)$.

(10) If A is $(Nv)B(v)$, then I satisfies A iff in every n distinct t-variants of I there is at least one which satisfies $B(t)$.

(11) If A is $(Mv)B(v)$, then I satisfies A iff there are n distinct t-variants of I, every one of which satisfies $B(t)$.

Notice that if D contains fewer than n elements, there are no n distinct t-variants of I. In that case $(Nv)B(v)$ is vacuously satisfied, and $(Mv)B(v)$ is obviously *not* satisfied.

A wff A is *valid* iff every interpretation satisfies A.

A wff A is *satisfiable* iff there is some interpretation that satisfies A.

A set of wffs Γ is *simultaneously satisfiable* iff there is some interpretation that satisfies every member of Γ.

A wff A is a *consequence* of a set of wffs Γ iff there is no interpretation that satisfies every member of Γ and does not satisfy A.

A sequent-expression $A_1, \ldots A_n \vdash B$ is *valid* iff B is a consequence of the set $\{A_1, \ldots A_n\}$.

Among the generalizations that can be established on the basis of the definitions are the following:

(1) If A is a consequence of Γ, and each wff in Γ is a consequence of Δ, then A is a consequence of Δ.

(2) A is a consequence of the empty set of sentences iff A is valid.

(3) B is a consequence of $\{A_1, A_2, \ldots A_n\}$ iff $(A_1 \And A_2 \And \ldots \And A_n) \rightarrow B$ is valid.

(4) If A is a member of Γ, then A is a consequence of Γ.

(5) If A is a consequence of Γ, and Γ is a subset of Δ, then A is a consequence of Δ.

(6) A is a consequence of $\Gamma U \{B\}$ iff $B \rightarrow A$ is a consequence of Γ.

(7) If A is a consequence of Γ, and every member of Γ is valid, then A is valid.

(8) A is valid iff $\sim A$ is not satisfiable.

(9) If I and I' have the same domain and agree in what they assign to all the terms and predicate letters occurring in A, then I satisfies A iff I' satisfies A.

(10) If A is like A' except for having the distinct terms $s_1, \ldots s_n$ wherever A' has the distinct terms $t_1, \ldots t_n$, and if I' is like I except for assigning to $t_1, \ldots t_n$ respectively what I assigns to $s_1, \ldots s_n$, then I satisfies A iff I' satisfies A'.

(11) Let $A(v)$ be a propositional function in v, and let $A(t)$ be the result of replacing all and only occurrences of v in $A(v)$ by t. Then $A(t)$ is a consequence of $(\forall v)A(v)$.

(12) Let $A(v)$ and $A(t)$ be as in (11). If $A(t)$ is a consequence of Γ, and t occurs neither in Γ nor in $A(v)$, then $(\forall v)A(v)$ is a consequence of Γ.

(13) Let $A(v)$ be a propositional function in v, and let $t_1, \ldots t_n$ be n distinct terms. Let $A(t_1), \ldots A(t_n)$ be the results of replacing all and only occurrences of v in $A(v)$ by $t_1, \ldots t_n$

respectively. Then $t_i \neq t_j \rightarrow \vee A(t_i)$ is a consequence of $(Nv)A(v)$.

(14) Let $A(v)$, $t_1, \ldots t_n$, $A(t_1), \ldots A(t_n)$ be as in (13). If $t_i \neq t_j \rightarrow \vee A(t_i)$ is a consequence of Γ, and no t_i occurs in $A(v)$, or in Γ, then $(Nv)A(v)$ is a consequence of Γ.

(15) Let I be an interpretation, and let $I'_1, \ldots I'_n$ be n distinct t-variants of I. Let J be like I except for assigning to $t_1, \ldots t_n$ what $I'_1, \ldots I'_n$ assign respectively to t. Then, if $A(v)$ is a propositional function in v that does not contain any of $t, t_1, \ldots t_n$, and $A(t_k)$ is the result of replacing all and only occurrences of v in $A(v)$ by t_k (which is one of $t_1, \ldots t_n$), then J satisfies $A(t_k)$ iff I'_k satisfies $A(t)$.

(16) If every element of the domain is assigned by an interpretation I to at least one term, then

  (i) I satisfies $(\forall v)A(v)$ iff for every term t, I satisfies $A(t)$;

  (ii) I satisfies $(\exists v)A(v)$ iff for at least one term t, I satisfies $A(t)$;

  (iii) I satisfies $(Nv)A(v)$ iff for every n distinct terms, $t_1, \ldots t_n$, I satisfies $t_i \neq t_j \rightarrow (\vee A(t_i))$;

  (iv) I satisfies $(Mv)A(v)$ iff for some n distinct terms $t_1, \ldots t_n$, I satisfies $t_i \neq t_j$ & $(\& A(t_i))$;

where (in (i) and (ii)) $A(t)$ is the result of replacing all and only occurrences of v in $A(v)$ by t_1 and (in (iii) and (iv)) $A(t_1)$, $A(t_2), \ldots A(t_n)$ are the results of replacing all and only occurrences of v in $A(v)$ by $t_1, t_2, \ldots t_n$ respectively.

Proofs of most of these generalizations are either immediate from the definitions, or very easy. Here proofs are given only for (9), (12) and (14). The proofs of (10) and (15) are closely analogous to that for (9). First we define the *degree* of a wff as the number of connectives or quantifiers occurring in it. Thus 'Fa' is of degree 0, '$(\forall x)Fx$' is of degree 1, '$\sim(\forall x)Fx$' is of degree 2, and '$\sim(\forall x)\sim Fx$' is of degree 3. The proof of (9) is then by mathematical induction on the degree of a wff A.

Proof of (9)

Let I and I' be two interpretations with the same domain, which agree in what they assign to all the terms and predicate letters occurring in a wff A. If A is of degree 0, it is an atomic sentence, and hence either a propositional variable, or one of the forms $t_1 = t_2$ or $Kt_1, \ldots t_m$. In any of the three cases, it is obvious that I satisfies A iff I' satisfies A, since I and I' agree on what they assign to a propositional variable, on what they assign to t_1 and t_2, and on what they assign to K and $t_1, \ldots t_m$.

Now suppose (9) holds for all wffs of degree n, and let A be a wff of degree $n+1$.

Case 1. A is some wff $\sim B$. B is therefore of degree n. By hypothesis, (9) holds for B. So I satisfies B iff I' satisfies B. Hence I does not satisfy B iff I' does not satisfy B. Therefore I satisfies $\sim B$ iff I' satisfies $\sim B$.

Case 2. A is $(B \vee C)$, where B and C are each at most of degree n. By hypothesis, I satisfies B iff I' satisfies B, and I satisfies C iff I' satisfies C. I satisfies $(B \vee C)$ iff I satisfies B or I satisfies C. Hence I satisfies $(B \vee C)$ iff I' satisfies $(B \vee C)$.

Case 3. A is $(B \& C)$, where B and C are each at most of degree n.

Case 4. A is $(B \rightarrow C)$, where B and C are each at most of degree n.

Proof of Cases 3 and 4 is similar to that of Case 2.

Case 5. A is $(\forall v)B(v)$. Let t be a term not occurring in $(\forall v)B(v)$, and let $B(t)$ be the result of replacing all and only occurrences of v in $B(v)$ by t. I satisfies $(\forall v)B(v)$ iff every t-variant of I satisfies $B(t)$. Suppose there is a t-variant J of I that does not satisfy $B(t)$. Then there is a t-variant J' of I', which assigns to t what J assigns to t, and J and J' agree in what they assign to all terms and predicate letters occurring in $B(t)$. But $B(t)$ is of degree n, so by hypothesis (9) holds for it. Hence J' does not satisfy $B(t)$. Similarly, if there is a t-variant of I' that does not satisfy $B(t)$, there is a t-variant of I that does not satisfy $B(t)$. So every t-variant of I satisfies $B(t)$ iff every t-variant of I' satisfies $B(t)$. Hence I satisfies $(\forall v)B(v)$ iff I' satisfies $(\forall v)B(v)$.

Case 6. A is $(Nv)B(v)$. Let t and $B(t)$ be as in Case 5. $B(t)$ is of degree n. If there are n distinct t-variants of I, none of which satisfy $B(t)$, there are n distinct t-variants of I', which assign to t what the n t-variants of I assign to t respectively, and by hypothesis, none of these t-variants of I' satisfy $B(t)$. So if in every n distinct t-variants of I' at least one satisfies $B(t)$, in every n distinct t-variants of I at least one satisfies $B(t)$. The argument is the same for the converse implication, which proves Case 6.

(10) and (15) can be proved by a closely similar inductive argument, on the degree of a wff A.

Proof of (12)

Suppose $A(t)$ is a consequence of Γ, and t occurs neither in Γ nor in $A(v)$ (where $A(v)$ is a propositional function from which $A(t)$ results by replacing v throughout by t). Let I be an interpretation that satisfies Γ. Let I' be a t-variant of I. By (9), I' satisfies Γ, and so I' satisfies $A(t)$. This holds for any t-variant of I, so I satisfies $(\forall v)A(v)$.

Proof of (14)

Let $A(v)$ be a propositional function in v, and let $A(t_1), \ldots A(t_n)$ be the results of replacing v in $A(v)$ by $t_1, \ldots t_n$ respectively. Suppose $t_i \neq t_j \rightarrow \vee A(t_i)$ is a consequence of Γ, and no t_i occurs in $A(v)$ or in Γ. Then let t be a term not occurring in Γ or in $t_i \neq t_j \rightarrow \vee A(t_i)$, and let $A(t)$ be the result of replacing v throughout $A(v)$ by t. Suppose I satisfies Γ.

Suppose there are n distinct t-variants of I, $I'_1, \ldots I'_n$, none of which satisfies $A(t)$. Let J be like I except for assigning to $t_1, \ldots t_n$ what $I'_1, \ldots I'_n$ assign to t respectively. By (9), J satisfies Γ. Hence it satisfies $t_1 \neq t_j \rightarrow \vee A(t_i)$. Since J assigns distinct elements to each of $t_1, \ldots t_n$, J satisfies $t_i \neq t_j$. Hence J satisfies $\vee A(t_i)$. Therefore, for some $k(1 \leq k \leq n)$, J satisfies $A(t_k)$. Hence, by (15), I'_k satisfies $A(t_k)$, contrary to hypothesis. Hence in any n distinct t-variants of I, at least one satisfies $t_i \neq t_j \rightarrow \vee A(t_i)$. Hence I satisfies $(Nv)A(v)$.

SOUNDNESS

To say that a system of rules is *sound* is to say that in any derivation conducted according to the rules, a wff appearing on the last line is a consequence of the assumptions on which it rests. The soundness of the rules of each system W_n is proved by induction on the length of a derivation. We prove (1) that any wff entered on the first line of a derivation is a consequence of the assumptions on which it rests, and (2), that any wff appearing on a later line is a consequence of the assumptions on which it rests, if all wffs occurring on earlier lines are consequences of the assumptions on which they rest.

(1) If a wff is entered on the first line of a derivation, it is entered by Rule A, resting upon itself. But any wff is a consequence of itself.

(2) It is easy to verify that if all wffs occurring on earlier lines are consequences of the assumptions on which they rest, and a wff is entered as a further line by one of the rules for the connectives, that wff is a consequence of the assumptions on which it rests. We therefore proceed at once to the rules for the quantifiers. Suppose then that all wffs occurring on earlier lines are consequences of the assumptions on which they rest. Then:

For UE. If a wff $A(t)$ is entered on a further line by UE, $(\forall v)A(v)$ occurs as an earlier line. $A(t)$ is a consequence of $(\forall v)A(v)$, by generalization (11). By hypothesis, $(\forall v)A(v)$ is a consequence of the assumptions on which it rests. $A(t)$ rests on the same assumptions. So $A(t)$ is a consequence of the assumptions on which it rests.

For UI. If a wff $(\forall v)A(v)$ is entered by UI, $A(t)$ occurs on an earlier line. Suppose $A(t)$ rests on assumptions Γ. Then t occurs neither in Γ nor in $A(v)$. By hypothesis $A(t)$ is a consequence of Γ. Hence, by generalization (12), $(\forall v)A(v)$ is a consequence of Γ.

For NE. If a wff $t_i \neq t_j \to \bigvee A(t_i)$ is entered by NE, $(Nv)A(v)$ occurs on an earlier line. $t_i \neq t_j \to \bigvee A(t_i)$ is a consequence of $(Nv)A(v)$, by (13). By hypothesis, $(Nv)A(v)$ is a consequence of the assumptions on which it rests. So $t_i \neq t_j \to \bigvee A(t_i)$ is a consequence of those same assumptions.

For NI. If a wff $(Nv)A(v)$ is entered by NI, $t_i \neq t_j \to \bigvee A(t_i)$ occurs on an earlier line, resting on assumptions Γ, where none of $t_1, \ldots t_n$ occur in Γ or in $A(v)$. By hypothesis $t_i \neq t_j \to \bigvee A(t_i)$ is a consequence of Γ. Hence, by generalization (14), $(Nv)A(v)$ is a consequence of Γ.

The soundness of the rules for identity can also easily be proved. Moreover, since any wff derived from given assumptions by the help of ME, MI, EI or EE can be derived without these rules, from the same assumptions, if Def $\exists$ and Def M are adopted; and since the rules of satisfaction justify Def $\exists$ and Def M, as the reader can verify, the rules ME, MI, EI and EE are sound if the other rules are sound. But the other rules *are* sound, as we have seen. Hence the system W_n is sound.

H Completeness

To say that a system of rules is complete is to say that if A is a consequence of Γ, then A is derivable from Γ, in accordance with the rules. To prove this, we need some definitions and preliminary lemmas.

A set of wffs Γ is said to be *consistent* iff no wff of the form $A \ \& \sim A$ is derivable from Γ. Note that B is derivable from Γ iff the set $\{\Gamma, \sim B\}$ is not consistent. For if B is derivable from Γ, $B \ \& \sim B$ is clearly derivable from $\{\Gamma, \sim B\}$, so $\{\Gamma, \sim B\}$ is not consistent. On the other hand if $\{\Gamma, \sim B\}$ is not consistent, any wff is derivable from it, so B is derivable from it.

Γ is said to be *maximal consistent* iff (1) Γ is consistent, (2) for any wff B that does not belong to Γ, the set $\{\Gamma, B\}$ is not consistent.

If Γ is maximal consistent, and A is any wff, then:

(1) A belongs to Γ iff $\sim A$ does not belong to Γ.

Proof

Suppose A belongs to Γ. Then, if $\sim A$ also belongs to Γ, $A \ \& \sim A$ is derivable from Γ. But this is impossible, since Γ is consistent. So $\sim A$ does not belong to Γ. Suppose $\sim A$ does not belong to Γ. Then $\{\Gamma, \sim A\}$ is not consistent, since Γ is maximal consistent. If A also does not belong to Γ, $\{\Gamma, A\}$ is not consistent. Then $A \ \& \sim A$ is derivable from Γ, which is impossible, since Γ is consistent. So A belongs to Γ.

(2) A belongs to Γ iff A is derivable from Γ.

Proof

If A belongs to Γ, then obviously A is derivable from Γ. If A does not belong to Γ, then $\sim A$ belongs to Γ (by (1)). Hence $\sim A$ is derivable from Γ. Since Γ is consistent, A is not derivable from Γ.

(3) $(A \lor B)$ belongs to Γ iff A belongs to Γ or B belongs to Γ.

(4) $(A \ \& B)$ belongs to Γ iff A belongs to Γ and B belongs to Γ.

(5) $(A \to B)$ belongs to Γ iff A does not belong to Γ or B belongs to Γ.

(3)–(5) follow easily from (1) and (2).

Γ is said to be *ω-complete* iff, if $(\exists v)A(v)$ belongs to Γ, then, for some term t, $A(t)$ belongs to Γ.

If Γ is maximal consistent and ω-complete, then (1)–(6) hold, and also

(7) $(\forall v)A(v)$ belongs to Γ iff for every t, $A(t)$ belongs to Γ.

(8) $(\exists v)A(v)$ belongs to Γ iff for some t, $A(t)$ belongs to Γ.

Proof of (7)

Suppose $(\forall v)A(v)$ belongs to Γ. Then, by *UE*, $A(t)$ is derivable from Γ, for every term t. Hence, by (2), $A(t)$ belongs to Γ. On the other hand, suppose that for every t, $A(t)$ belongs to Γ. Now suppose $(\forall v)A(v)$ does *not* belong to Γ. Then $\sim(\forall v)A(v)$ belongs to Γ (by (1)). By (2), $(\exists v)\sim A(v)$ belongs to Γ. Then, since Γ is ω-complete, $\sim A(t)$ belongs to Γ, for some t. But since $A(t)$ is by hypothesis a member of Γ, Γ is not consistent, contrary to hypothesis. So $(\forall v)A(v)$ belongs to Γ.

Proof of (8)

If $(\exists v)A(v)$ belongs to Γ, then, for some t, $A(t)$ belongs to Γ, since Γ is ω-complete. If, for some t, $A(t)$ belongs to Γ, then $(\exists v)A(v)$ is derivable from Γ, and hence belongs to Γ.

Γ is said to be *M-ω-complete* iff, if $(Mv)A(v)$ belongs to Γ, then, for some n distinct terms $t_1, \ldots t_n$, $t_i \neq t_j$ & $(\&\ A(t_i))$ belongs to Γ.

If Γ is maximal consistent, ω-complete and *M-ω-complete*, then

(9) $(Nv)A(v)$ belongs to Γ iff for every n distinct terms $t_1, \ldots t_n$, $t_i \neq t_j \rightarrow \vee A(t_i)$ belongs to Γ.

Proof of (9)

Suppose $(Nv)A(v)$ belongs to Γ. Then, by *NE*, $t_i \neq t_j \rightarrow \vee A(t_i)$ is derivable from Γ for any n terms $t_1, \ldots t_n$. Hence, by (2), $t_i \neq t_j \rightarrow \vee A(t_i)$ belongs to Γ. Now suppose that for every n terms, $t_i \neq t_j \rightarrow \vee A(t_i)$ belongs to Γ. Suppose that $(Nv)A(v)$ does not belong to Γ. Then $\sim(Nv)A(v)$ belongs to Γ, by (1). By (2), $(Mv)\sim A(v)$ belongs to Γ. Then, since Γ is *M-ω-complete*, for some $t_1, \ldots t_n$, $t_i \neq t_j$ & $(\&\ \sim A(t_i))$ belongs to Γ. But since $t_i \neq t_j \rightarrow \vee A(t_i)$ belongs to Γ, Γ is not consistent, contrary to hypothesis. So $(Nv)A(v)$ belongs to Γ.

(10) $(Mv)A(v)$ belongs to Γ iff for some n-terms $t_1, \ldots t_n$, $t_i \neq t_j$ & $(\&\ A(t_i))$ belongs to Γ.

Proof of (10)

If $(Mv)A(v)$ belongs to Γ, $t_i \neq t_j$ & $(\&\ A(t_i))$ belongs to Γ, for some $t_1, \ldots t_n$, since Γ is *M-ω-complete*. If $t_i \neq t_j$ & $(\&\ A(t_i))$

belongs to Γ, then $(Mv)A(v)$ is derivable from Γ, by MI. Hence $(Mv)A(v)$ belongs to Γ, by (2).

Lemma I. If Γ is consistent, then it is simultaneously satisfiable.

Once that Lemma I is proved, completeness can be quickly established as follows: Suppose that A is a consequence of Γ. Then $\{\Gamma, \sim A\}$ is not simultaneously satisfiable. Hence, by Lemma I, $\{\Gamma, \sim A\}$ is not consistent. Therefore, A is derivable from Γ. Thus, if A is a consequence of Γ, A is derivable from Γ, which proves completeness.

For the proof of Lemma I, we adapt Benson Mates's presentation, in his *Elementary Logic* (pp. 138–40), of Henkin's completeness proof for classical predicate logic. We prove Lemma I by proving a special case, Lemma I′, from which Lemma I follows.

Lemma I′. If Γ is consistent, and all subscripts of terms in Γ are even, then Γ is simultaneously satisfiable. (The unsubscripted terms 'a', 'b', 'c', 'd', are treated as if they had the subscript '0'.)

The restriction about subscripts guarantees that there are infinitely many terms not occurring in Γ, which turns out to be useful. It is clear that Lemma I follows from Lemma I′. For given a set Γ, doubling all the subscripts of terms occurring in Γ cannot affect the consistency or inconsistency of Γ.

Lemma I′ follows immediately from the two further Lemmas.

Lemma II. If Γ satisfies the antecedent of Lemma I′, then there exists a set Δ that includes Γ and is maximal consistent, ω-complete and M-ω-complete.

Lemma III. Every maximal consistent, ω-complete and M-ω-complete set of wffs is simultaneously satisfiable.

Proof of Lemma II

All the wffs of W_n can be arranged in an infinite list
$$A_1, A_2, \ldots A_n, \ldots$$
with the properties

(a) Each wff of W_n occurs at least once in the list.

(b) For each wff of the form $(\exists v)A(v)$, there is at least once i such that A_i is $(\exists v)A(v)$, and $i+1$ is $A(t)$, where t is a term which occurs neither in Γ, nor in any of $A_1, A_2, \ldots A_i$.

(c) For each wff of the form $(Mv)A(v)$, there is at least one

j such that A_j is $(Mv)A(v)$, and $j+1$ is $t_k \neq t_i$ & ($\&$ $A(t_k)$), where $t_1, \ldots t_n$ are n distinct terms none of which occur either in Γ or in any of $A_1, A_2, \ldots A_j$.

It is here that the restriction on subscripts is useful. When we list the wffs, whenever we come to $(\exists v)A(v)$ we wish to throw in $A(t)$, where t is a term not previously occurring in the list or in Γ, and whenever we come to $(Mv)A(v)$ in the list we wish to throw in $t_k \neq t_i$ & ($\&$ $A(t_k)$), where $t_1, \ldots t_n$ are terms not previously occurring either in the list or in Γ. To be assured that this can be done, there must be assurance that there are always fresh terms available. This is ensured by the restriction that Γ contains none of the infinitely many odd-numbered terms. The list is then constructed by first enumerating the symbols of W_n; second, by using this enumeration to obtain a list of the *formulae*; third, by striking out from this list all formulae that are not well-formed; and finally, by going through the resulting reduced list and adding wffs so as to satisfy (b) and (c). We do not go into further details of this process.

Now, using this list, we construct an infinite sequence of sets of wffs

$$\Delta_0, \Delta_1, \ldots \Delta_n, \ldots$$

as follows:

Δ_0 is Γ. Δ_1 is $\{\Delta_0, A_1\}$ if $\{\Delta_0, A_1\}$ is consistent. Otherwise Δ_1 is Δ_0. In general, Δ_n is $\{\Delta_{n-1}, A_n\}$ if $\{\Delta_{n-1}, A_n\}$ is consistent, and otherwise Δ_n is Δ_{n-1}.

Let Δ be the union of all the sets of $\Delta_0, \Delta_1, \ldots \Delta_n, \ldots$, so that a wff B belongs to Δ iff B belongs to Δ_i, for at least one i. We prove that Δ includes Γ, is maximal consistent, is ω-complete, and M-ω-complete.

(1) Δ obviously includes Γ.

(2) Δ is consistent. By the construction, each Δ_i is consistent, since Γ is consistent. If Δ is not consistent, A & $\sim A$ is derivable from Δ. Then A & $\sim A$ is derivable from some finite subset Δ' of Δ, since any derivation is of finite length, and hence contains only finitely many assumptions. But Δ' must be included in Δ_j, for some j. For if A_j is the wff in Δ' that occurs latest in the list of all the wffs, Δ' is included in Δ_j. But then A & $\sim A$ is derivable from Δ_j, which is therefore not consistent. This is contrary to hypothesis. Hence Δ is consistent.

E

(3) Δ is maximal consistent. Let A be a wff that is not a member of Δ. A is A_i, for some i. Since A does not belong to Δ, it does not belong to Δ_i. Hence $\{\Delta_{i-1}, A\}$ is not consistent. Therefore $\{\Delta, A\}$ is not consistent. Hence Δ is maximal consistent.

(4) Δ is ω-complete. Suppose that $(\exists v)A(v)$ belongs to Δ. Then, for some i, $(\exists v)A(v)$ is A_i, and A_{i+1} is $A(t)$, where t occurs neither in Γ nor in any earlier wff in the list. $(\exists v)A(v)$ is a member of Δ_i, by the construction of Δ. Now suppose $A(t)$ does not belong to Δ_{i+1}. Then $\{\Delta_i, A(t)\}$ is not consistent. Hence $\sim A(t)$ is derivable from Δ_i. Hence, by UI, $(\forall v)\sim A(v)$ is derivable from Δ_i, since t does not occur in Δ_i. But since $(\exists v)A(v)$ belongs to Δ_i, $\sim(\forall v)\sim A(v)$ is derivable from Δ_i. So Δ_i is not consistent, contrary to hypothesis.

(5) Δ is M-ω-complete. Suppose that $(Mv)A(v)$ belongs to Δ. For some i, $(Mv)A(v)$ is A_i, and A_{i+1} is $t_k \neq t_i \mathbin{\&} (\mathbin{\&} A(t_k))$, where none of $t_1, \ldots t_n$ occur in Γ or in any earlier wff in the list. $(Mv)A(v)$ belongs to Δ_i. Suppose $t_k \neq t_i \mathbin{\&} (\mathbin{\&} A(t_k))$ does not belong to Δ_{i+1}. Then $\{\Delta_i, t_k \neq t_i \mathbin{\&} (\mathbin{\&} A(t_k))\}$ is not consistent. Hence $t_k \neq t_i \rightarrow \bigvee \sim A(t_k)$ is derivable from Δ_i. Hence, by NI, $(Nv)\sim A(v)$ is derivable from Δ_i. But since $(Mv)A(v)$ belongs to Δ_i, $\sim(Nv)\sim A(v)$ is derivable from Δ_i. So Δ_i is not consistent, contrary to hypothesis.

Thus the proof of Lemma II is now complete.

Proof of Lemma III

Let Δ be a set of wffs that is maximal consistent, ω-complete and M-ω-complete. We specify an interpretation, and prove that it simultaneously satisfies Δ.

The domain of the interpretation I is as follows. We divide the terms of W_n into classes in such a way that t_1 and t_2 belong to the same class iff the wff $t_1 = t_2$ belongs to Δ. We write '$[t]$' for 'the class to which t belongs'. The domain of I is then the set of all these classes.

Now I assigns to each term t the class $[t]$. And I assigns to each predicate letter K of degree n the set of n-tuples $<[t_1]$, $[t_2], \ldots [t_n]>$, which are such that $Kt_1, \ldots t_n$ belongs to Δ. Note that if $[s_1] = [t_1], [s_2] = [t_2], \ldots [s_n] = [t_n]$, then $Ks_1, \ldots s_n$ belongs to Δ iff $Kt_1, \ldots t_n$ belongs to Δ. For if $[s_i] = [t_i]$, then $s_i = t_i$ belongs to Δ. Hence if $Ks_1, \ldots s_n$ belongs to Δ, $Kt_1, \ldots t_n$ is derivable from Δ, by the rule IE, and so $Kt_1, \ldots t_n$ belongs to Δ. Similarly, if $Kt_1, \ldots t_n$ belongs to Δ, $Ks_1, \ldots s_n$ belongs to Δ.

It is clear that I assigns the identity-relation to ' $=$ '. Finally, I assigns the value T to a propositional variable if it belongs to Δ, and otherwise F.

We now prove that a wff A belongs to Δ iff I satisfies A. As before, we define the *degree* of a wff as the number of connectives or quantifiers it contains. The proof is by induction on the degree of a wff.

Suppose A is a wff of degree 0. Then A is an atomic sentence. It is clear that I satisfies A iff A belongs to Δ.

Now suppose that, for all wffs A of degree n, I satisfies A iff A belongs to Δ. Let B be a wff of degree $n+1$.

Case 1. B is $\sim C$, where C is of degree n. By hypothesis, I satisfies C iff C belongs to Δ. I satisfies $\sim C$ iff I does not satisfy C. Hence I satisfies $\sim C$ iff C does not belong to Δ. But C does not belong to Δ iff $\sim C$ belongs to Δ, since Δ is maximal consistent. Hence I satisfies $\sim C$ iff $\sim C$ belongs to Δ.

Case 2. B is $(C \vee D)$, where C and D are each at most of degree n. I satisfies $(C \vee D)$ iff I satisfies C or I satisfies D. By hypothesis, I satisfies C or satisfies D iff either C or D belongs to Δ. C or D belong to Δ iff $(C \vee D)$ belongs to Δ, since Δ is maximal consistent. Hence I satisfies $(C \vee D)$ iff $(C \vee D)$ belongs to Δ.

Case 3. B is $(C \mathbin{\&} D)$.

Case 4. B is $(C \to D)$.

Proof of Cases 3 and 4 is similar to that of Case 2.

Case 5. B is $(\forall v)C(v)$. Then, for any term t, $C(t)$ is of degree n. Since every element of the domain is assigned to at least one term, I satisfies $(\forall v)C(v)$ iff, for every term t, I satisfies $C(t)$. By hypothesis, for every term t, I satisfies $C(t)$ iff $C(t)$ belongs to Δ. Since Δ is maximal consistent and ω-complete, $C(t)$ belongs to Δ, for every term t, iff $(\forall v)C(v)$ belongs to Δ. Hence I satisfies $(\forall v)C(v)$ iff $(\forall v)C(v)$ belongs to Δ.

Case 6. B is $(\exists v)C(v)$. I satisfies $(\exists v)C(v)$ iff for some t, I satisfies $C(t)$, since every element is assigned to at least one term. By hypothesis, I satisfies $C(t)$ iff $C(t)$ belongs to Δ. Since Δ is ω-complete, $C(t)$ belongs to Δ, for some t, iff $(\exists v)C(v)$ belongs to Δ. Hence I satisfies $(\exists v)C(v)$ iff $(\exists v)C(v)$ belongs to Δ.

Case 7. B is $(Nv)C(v)$. The proof is similar to that of Case 5 using the M-ω-completeness of Δ instead of its ω-completeness.

Case 8. B is $(Mv)C(v)$. The proof is similar to that of Case 6, using M-ω-completeness instead of ω-completeness.

This completes the proof that, for all A, I satisfies A iff A belongs to Δ. Hence I satisfies simultaneously every wff in Δ, which proves Lemma III. Lemma I' follows immediately from Lemmas II and III. Lemma I follows from Lemma I', as we have pointed out. And completeness follows from Lemma I in the way we have shown.

Since the interpretation I specifies a denumerable domain, we have also proved that any consistent set of wffs is simultaneously satisfiable in a denumerable domain. Since the rules are sound, a simultaneously satisfiable set of wffs is consistent. Consequently, we have the Lowenheim-Skolem theorem for the logic of plurality, that any simultaneously satisfiable set of wffs is simultaneously satisfiable in a denumerable domain.

CHAPTER THREE

The stronger systems

The systems W_n contain all those principles that are valid in every non-empty domain. The stronger systems we are now to glance at contain those principles that are valid only in domains with *many* members. That is to say, we now investigate the second possible assumption about domains of discourse mentioned at the end of Chapter 1. It may be thought more natural, in dealing with the logic of plurality, to assume that the domain of discourse contains many members. This intuitive feeling is supported by the greater simplicity of the rules of the systems we shall now present.

A   The primitive basis of S_n

Let n be some fixed positive integer greater than one. Then the vocabulary of the system S_n consists of: the proper names, individual variables, predicate letters of degree n, for each n, the connectives, the quantifier-symbols '$\forall$', 'N', and the brackets '(', ')'. In addition, an infinite list of n-tuples of *constants*,

$$e_{01}, \ldots e_{0n}; \ e_{11}, \ldots e_{1n}; \ e_{m1}, \ldots e_{mn}.$$

We adopt Def $\exists$ and Def M, to define the existential and plural quantifiers respectively. The identity-sign is not included among the primitive signs, for we shall here treat of the calculus of plurality without identity, for the sake of simplicity.

A *term* is now defined to be either a proper name or a constant. As before, a predicate letter of degree zero is called a propositional variable, and the same symbols 'P', 'Q', 'R', etc., are used for these as in Chapter 2. And a formula is any finite sequence of symbols. A formula is an *atomic sentence* iff it is either a propositional variable or a predicate letter of degree n followed immediately by a string of n terms.

Wffs are defined as follows:

(1) If A is an atomic sentence, A is a wff.

(2) If A is a wff, $\sim A$ is a wff.

(3) If A and B are wffs, then $(A \lor B)$, $(A \& B)$, $(A \rightarrow B)$ are wffs.

(4) Let $A(t)$ be a wff containing the term t, and let v be a variable not occurring in $A(t)$. Let $A(v)$ be a formula resulting from replacing at least one occurrence of t in $A(t)$ by v. Then $(\forall v)A(v)$, and $(Nv)A(v)$ are wffs.

(5) If a formula is not a wff in virtue of 1–4, then it is not a wff.

In addition to the conventions of Chapter 2 regarding syntactic variables, we use $g_1, \ldots g_n$ as variables ranging over constants only.

Derivations and proofs are written as in Chapter 2. As rules of derivation, we adopt the same rules for the connectives, and also the rules UE and UI, with the sole difference that the terms now include the constants as well as the proper names. In addition, we add two rules corresponding to NE and NI, which will bear the same names.

NE

Let $A(v)$ be a propositional function in v, and let $g_1, \ldots g_n$ be one of the n-tuples of *constants*. Let $A(g_1)$, $A(g_2)$, $\ldots A(g_n)$ be the results of replacing all and only occurrences of v in $A(v)$ by g_1, g_2, $\ldots g_n$ respectively. Then given $(Nv)A(v)$, the wff $\lor A(g_i)$ may be entered as a further line, depending on the same assumptions as $(Nv)A(v)$.

NI

Let $A(t)$ be a wff containing a term t, let v be a variable not occurring in $A(t)$, and let $g_1, g_2, \ldots g_n$ be one of the n-tuples of constants. Let $A(g_1)$, $A(g_2)$, $\ldots A(g_n)$ be the results of replacing all and only occurrences of t in $A(t)$ by $g_1, g_2, \ldots g_n$ respectively. Let $A(v)$ be the propositional function in v that results from replacing all and only occurrences of t in $A(t)$ by v. Then given $\lor A(g_i)$, $(Nv)A(v)$ may be entered as a further line, depending on the same assumptions as $\lor A(g_i)$, provided that none of $g_1, \ldots g_n$ occurs in any assumption upon which $\lor A(g_i)$ rests.

B The interpretation of S_n

An interpretation of S_n consists of a domain containing at least n members, together with a function that assigns to every propositional variable one of the two truth-values, to every proper name an element from the domain, to every constant an element from the domain, in such a way that *in any n-tuple of constants, no two constants are assigned the same element*, and to every predicate letter of degree n a set of n-tuples from the domain.

Now the essential difference between S_n and W_n becomes clear. The domain of an interpretation for W_n had only to be non-empty. The domain for S_n contains at least n members. Any n-tuple of constants are all assigned distinct elements from the domain in S_n. This is made possible by the provision that the domain must contain n elements. The purpose of the constants, so interpreted, is to enable us to dispense with the cumbersome identity-formulae in the formulation of NE and NI, and hence in the work of conducting derivations. Constants could not be used in this way in W_n, since in domains with fewer than n elements, some of them would have to remain uninterpreted.

The rules of satisfaction are the same as in Chapter 2 for the atomic sentences, the rules for the connectives, and the universal quantifier. For a wff of the form $(Nv)A(v)$, we say that an interpretation I satisfies $(Nv)A(v)$ iff in every n distinct t-variants of I, at least one satisfies $A(t)$. Now since the domain contains at least n elements, there always exist at least n distinct t-variants of I. So in S_n, $(Nv)A(v)$ is never vacuously satisfied.

The soundness of the rules of S_n can be established by arguments similar to those of Chapter 2. We do not here go into the metatheory in detail, but proceed to illustrate the working of the rules. The rules EI and EE can be obtained as derived rules, as can the rules ME and MI (which have certain obvious differences from the corresponding rules for W_n, namely that the identity-formulae are dispensed with, and the rules are formulated in terms of constants rather than proper names).

C   Sequents of S_n

It is clear that the sequents of W_n that do not contain the identity-sign are provable in S_n also. We merely give a few examples.

$(SS1)$ $(Mx)(Fx \& Gx) \vdash (Mx)Fx \& (Mx)Gx$

Proof

1	(1) $(Mx)(Fx \& Gx)$	A
2	(2) $\& (Fe_i \& Ge_i)$	A
2	(3) $\& Fe_i$	(2) $\& E, \& I$
2	(4) $\& Ge_i$	(2) $\& E, \& I$
2	(5) $(Mx)Fx$	(3) MI
2	(6) $(Mx)Gx$	(4) MI
2	(7) $(Mx)Fx \& (Mx)Gx$	(5),(6) $\& I$
1	(8) $(Mx)Fx \& (Mx)Gx$	(1),(2),(7) ME

(SS2) $(Nx)(Fx \& Gx) \vdash (Nx)Fx \& (Nx)Gx$

Proof

1	(1) $(Nx)(Fx \& Gx)$	A
1	(2) $\vee (Fe_i \& Ge_i)$	(1) NE
1	(3) $\vee Fe_i$	(2) RC
1	(4) $\vee Ge_i$	(2) RC
1	(5) $(Nx)Fx$	(3) NI
1	(6) $(Nx)Gx$	(4) NI
1	(7) $(Nx)Fx \& (Nx)Gx$	(5),(6) $\& I$

(SS3) $(Nx)(Fx \to P) \vdash (Mx)Fx \to P$

Proof

1	(1) $(Nx)(Fx \to P)$	A
2	(2) $(Mx)Fx$	A
3	(3) $\& Fe_i$	A
1	(4) $\vee (Fe_i \to P)$	(1) NE
1,3	(5) P	(3),(4) RC
1,2	(6) P	(2),(3),(5) ME
1	(7) $(Mx)Fx \to P$	(2),(6) CP

Certain sequent-expressions that are not provable in W_n become provable in S_n. For instance

(SS4) $(\forall x)Fx \vdash (Mx)Fx$

(SS4) is not valid in W_n, since the domain of discourse may contain fewer than n members. But it *is* valid in S_n.

Proof of (SS4)

1	(1) $(\forall x)Fx$	A
1	(2) $\& Fe_i$	(1) $UE_n, \& I$
1	(3) $(Mx)Fx$	(2) MI

Also provable in S_n, but not in W_n, is

(SS5) $(Nx)Fx \vdash (\exists x)Fx$

	Proof	
1	(1) $(Nx)Fx$	A
1	(2) $\vee Fe_i$	(1) NE
1	(3) $(\exists x)Fx$	(2) $\vee E, EI$

In W_n it is possible to prove both

$$(\forall x)Fx, (Mx)Gx \vdash (Mx)Fx$$

and

$$(Nx)Fx, (Mx)Gx \vdash (\exists x)Fx.$$

And indeed the main difference between the S_n and the W_n is that in S_n certain sequents are provable in a stronger form, without an extra assumption of the form $(Mv)A(v)$. A closely related difference is that in S_n, '$\sim(Nx)(Fx \& \sim Fx)$' is a theorem, but that formula is not valid in W_n. These features give S_n a greater naturalness than W_n, at the expense of diverging from the assumption about the domain of discourse generally made in classical logic.

The system S_n can be thought of as the logic of the numerical quantifier '$(\exists_n v)A(v)$', 'there are at least n objects such that A', where the underlying domain is assumed to have at least n members. It becomes such a logic upon replacing (Mv) by $(\exists_n v)$, and (Nv) by $\sim(\exists_n v)\sim$. The principles that are common to all S_n are the principles that hold for $(\exists_n v)$, provided that the domain contains n members, irrespective of the value of n.

D   'A few'

In Chapter 1 we distinguished 'few' from 'a few'. 'There are few Fs' was said to be equivalent to 'There are not many Fs', and hence to follow from 'There are no Fs'. No separate treatment of 'few' was therefore necessary. It was taken to be covered by the treatment of 'many' together with 'not'. This is so whether we are working in the systems W_n or in S_n. But 'There are *a* few Fs' is different. So far from following from 'There are no Fs' is it that it is incompatible with the latter, and actually entails 'There is at least one F'. We also briefly distinguished 'Nearly everything is F' from 'All but a few things are F'.

While 'Nearly everything is F' follows from 'Everything is F', 'All but a few things are F' actually entails that *not* everything is F. The problem now is how to treat 'a few' and 'all but a few', since this is not obvious from what has been said so far.

There seem to be two possible methods. The first method treats 'a few' and 'all but a few' within the framework of our existing apparatus, by *defining* 'a few' as 'some, but not many'. Thus if we introduce a new quantifier $(Fv)A(v)$, to be read 'A few things are A', it can be defined as follows:

$$\text{Def } F\colon (Fv)A(v) = (\exists v)A(v) \;\&\; \sim(Mv)A(v).$$

Then $\vdash \sim(Fv)A(v) \leftrightarrow ((\forall v)\sim A(v) \vee (Mv)A(v))$.
Thus it is false that a few things are A iff either nothing has A or many things do.

Let us introduce a further quantifier $(\daleth v)A(v)$, to be read 'All but a few things are A'. Then we can give the definition

$$\text{Def } \daleth\colon (\daleth v)A(v) = (Nv)A(v) \;\&\; \sim(\forall v)A(v).$$

Consequently, it is false that all but a few things are A iff either many things are not A or everything is A, i.e.

$$\vdash \sim(\daleth v)A(v) \leftrightarrow ((\forall v)A(v) \vee (Mv)\sim A(v)).$$

But now it is clear that one of the new quantifiers can be defined in terms of the other. For

$$\vdash \sim(Fv)\sim A(v) \leftrightarrow ((\forall v)A(v) \vee (Mv)\sim A(v)).$$

Hence

$$\vdash \sim(\daleth v)A(v) \leftrightarrow \sim(Fv)\sim A(v).$$

Similarly

$$\vdash \sim(\daleth v)\sim A(v) \leftrightarrow \sim(Fv)A(v).$$

We could thus replace Def F by

$$\text{Def } 2F\colon (Fv)A(v) = (\daleth v)\sim A(v)$$

or replace Def $\daleth$ by

$$\text{Def } 2\daleth\colon (\daleth v)A(v) = (Fv)\sim A(v).$$

This method is highly convenient, but it may be thought unsatisfactory, since by this analysis, if there is exactly one F, it follows that there are a few Fs. And it may be thought that for there to be a few Fs there must be more than one. According to this line of thought 'a few' would be a quantifier lying some-where between 'many' and 'some'. We can do justice to this idea as follows.

Suppose we are working in W_n. We introduce a new *primitive* sign 'f', as our quantifier-sign for 'a few'. We let m be some positive integer strictly less than n and strictly greater than 1 (hence this method works only from systems W_3 upwards). We introduce a new rule of satisfaction, namely

(RF) An interpretation I satisfies $(fv)A(v)$ iff there are at least m distinct t-variants of I, all of which satisfy $A(t)$.

Finally, as rules of derivation, we have fE and fI, which are exactly analogous to ME and MI, with 'f' in place of 'M', and 'm' in place of 'n'.

According to the first method, 'a few' means 'at least one and at most a few', but according to the second method, 'a few' means 'at least a few'. The two methods are not incompatible. If we adopt both, then 'there are exactly a few' can be defined as $(fv)A(v)$ & $(Fv)A(v)$. But '$(Fv)A(v)$' is unnecessary for defining 'exactly a few'. The latter can equally be defined as $(fv)A(v)$ & $\sim(Mv)A(v)$.

The new quantifier (fv) has another use. It enables us to define the quantifier 'Very nearly all', or (Vv), by

$$\text{Def V: } (Vv)A(v) = \sim(fv)\sim A(v).$$

Thus I satisfies $(Vv)A(v)$ iff in every m distinct t-variants of I, at least one satisfies $A(t)$. Since m is less than n, we have

$$(Vv)A(v) \vdash (Nv)A(v).$$

Now 'All but at most a very few' might be expressed as

$$(Vv)A(v) \; \& \; (\exists v)\sim A(v).$$

These suggestions should suffice to show that our methods are sufficiently flexible to allow the treatment of several interesting quantifiers other than the ones we have studied in detail.

E A yet stronger system

A notable absentee from both the W-systems and the S-systems is the sequent-expression

$$(Nx)Fx \vdash (Mx)Fx$$

Indeed, this expression is clearly invalid, even in the S-systems. For let there be given a domain of exactly n members, all but one of which have the property F. Then '$(Nx)Fx$' is true, and '$(Mx)Fx$' is false. In spite of this, the above expression has a certain intuitive plausibility. This plausibility derives from the following fact: the above expression is valid in all domains that are sufficiently large in relation to n. This fact can be clarified as follows: if nearly everything is F, then at most $n-1$ things are not F. Hence if there are at least $2n-1$ things in the domain, many things have F.

Let us therefore get a glimpse of how to formulate systems that assume that the underlying domain has at least $2n-1$ members. We call these the systems T_n. For the systems T_n we introduce special constants, as we did in S_n, but in a different way. In T_n, the constants come, not in n-tuples, but in $(2n-1)$-tuples. Thus the vocabulary includes, for each k, a set of constants

$$e_{k1}, \; e_{k2}, \; \ldots \; e_{k2n-1}.$$

An interpretation I assigns distinct elements to distinct constants belonging to the same $(2n-1)$-tuple. This is always possible, since the domain contains at least $2n-1$ elements. Now let $c_1, \ldots c_n$ be a *combination* of n distinct constants from a given $(2n-1)$-tuple of special constants. And suppose that for any such combination, at least one of $A(c_1), A(c_2), \ldots A(c_n)$ is satisfied by I. Then I satisfies $(Mv)A(v)$. This is justified by the following argument. Take some n distinct constants $c_1, \ldots c_n$, from a $(2n-1)$-tuple. At least one, call it e_1, is such that I satisfies $A(e_1)$, by hypothesis. Now take an n-tuple of constants, not including e_1. Select one, call it e_2, such that I satisfies $A(e_2)$. Progressively take further n-tuples, not including any e_i previously selected as being such that I satisfies $A(e_i)$. This yields a sequence $A(e_1), A(e_2), \ldots A(e_n)$, all of which are satisfied by I, and where $e_1, e_2, \ldots e_n$ are distinct constants from the same $(2n-1)$-tuple. Hence I satisfies $(Mv)A(v)$, since $e_1, \ldots e_n$ are assigned distinct elements from the domain. The rule NE becomes

(NE_T) Let $A(v)$ be a propositional function in v, and let $g_1, \ldots g_n$ be n distinct constants from some $(2n-1)$-tuple of constants. Let $A(g_1), \ldots A(g_n)$ be the results of replacing v in

$A(v)$ by $g_1, \ldots g_n$ respectively. Then given $(Nv)A(v)$, $\bigvee A(g_i)$ may be entered as a further line, depending on the same assumptions.

The other rules differ from their counterparts in the S-systems in an analogous way.

Now, as an example, we prove '$(Nx)Fx \vdash (Mx)Fx$' in T_2 (in a somewhat abbreviated form).

$(ST1)$ $(Nx)Fx \vdash (Mx)Fx$

1	(1) $(Nx)Fx$	A
1	(2) $Fe_1 \lor Fe_2$	(1) NE
1	(3) $Fe_1 \lor Fe_3$	(1) NE
1	(4) $Fe_2 \lor Fe_3$	(1) NE
1	(5) $(Fe_1 \lor Fe_2) \;\&\; (Fe_1 \lor Fe_3) \;\&\; (Fe_2 \lor Fe_3)$	
		(2),(3),(4) $\&\, I$
1	(6) $(Fe_1 \;\&\; Fe_2) \lor (Fe_1 \;\&\; Fe_3) \lor (Fe_2 \;\&\; Fe_3)$	(5) RC
7	(7) $Fe_1 \;\&\; Fe_2$	A
7	(8) $(Mx)Fx$	(7) MI
9	(9) $Fe_1 \;\&\; Fe_3$	A
9	(10) $(Mx)Fx$	(9) MI
11	(11) $Fe_2 \;\&\; Fe_3$	A
11	(12) $(Mx)Fx$	(11) MI
1	(13) $(Mx)Fx$	(7)–(12) $\lor E$

Explanation: In T_2, $n = 2$. Hence a $(2n-1)$-tuple contains three constants. At (2), (3) and (4) we take all possible two-membered disjunctions from such a $(2n-1)$-tuple. At (6) we distribute, to form a disjunction of two-membered conjunctions. The rest of the proof is obvious.

It is clear that a similar proof can be carried out in any T_n. If we adopt the quantifiers (fv) and (Vv) in T_n, with appropriate rules, we can show that all the various quantifiers are ordered in strength as follows:

$$(\forall v)A(v) \vdash (Vv)A(v) \vdash (Nv)A(v)$$

$$\vdash (Mv)A(v) \vdash (fv)A(v) \vdash (\exists v)A(v).$$

The systems T_n are thus perhaps the most natural of all. Unhappily, they are rather complicated to work with.

CHAPTER FOUR

Further refinement: the attributive systems W^k, S^k, and T^k

The three systems, W, S and T all have their merits, and each of them can be used to represent formally a variety of patterns of informal argument. The system T, being the strongest, can represent the largest set of informal arguments, and is perhaps the most natural of the three. It is also the system that exploits most fully the analogies between the universal and the pen-universal quantifiers, and between the existential and the plural quantifiers. Another point of interest concerning it is that where the domain of discourse is understood to contain *exactly* $2n-1$ elements, the quantifier '(Mx)' can justly be read as 'most'. For most things have a property F, iff more than half have that property. In order that '$(Mx)Fx$' be true in a domain of exactly $2n-1$ things, n elements (at least) in that domain must have F. But n divided by $2n-1$ is always greater than a half. So if '$(Mx)Fx$' is true in such a domain, most things in that domain have F. Moreover, if most things have F, then more than half have F. But n is the least number which is more than half of $2n-1$. So '$(Mx)Fx$' will be true. Thus in such a domain 'most' and '(Mx)' are equivalent. Notice also that in a domain of exactly $2n-1$ elements, '(Mx)' and '(Nx)' are equivalent, i.e. 'not most not' is equivalent to 'most'. This is ensured by the fact that $2n-1$ is always an odd number. Thus T has a certain versatility, and points up an interesting relation between 'many' and 'most'. The equivalence of '(Mx)' and '(Nx)' in a domain with exactly $2n-1$ members is the analogue of the fact that in domains with exactly *one* member, the universal and existential quantifiers are equivalent. Indeed, this emerges from the consideration of T quite naturally, if n is taken as one: for then $2n-1$ is also one; '(Nx)' and '(Mx)' collapse into the universal and existential quantifiers respectively and these then collapse into each other. However, in general, '(Mx)' cannot be read as 'most' in T, since the interpretation of T requires only that there be *at least* $2n - 1$ elements in the domain. For 'most', we need the stronger requirement that the number be exactly $2n - 1$.

Together with what has gone before, these points are substantial evidence of the flexibility of the present methods. By varying the minimal size of manifolds, and by varying the minimum size of domains, or by fixing also a maximal size on a domain, it is possible to formalize a host of interesting quantifiers, and to point up the links between them.

But now it is time to show that there are very important limitations on the applicability of any of the systems to informal argument. Let us consider the formula '$(Mx)(Fx \ \& \ Gx)$', and the informal sentence-form 'Many Fs are Gs'. There is good reason to believe (1) that if 'Many Fs are Gs' is to be formally represented within T, it must be by '$(Mx)(Fx \ \& \ Gx)$', and (2) that such a representation is admissible only in the context of a limited class of arguments. Consider the following argument:

(*A*1) All Quakers are Christians. Many Quakers hate churches. So many Christians hate churches.

(*A*1) may well be judged to be invalid. However, suppose it is represented by

(*B*1) $(\forall x) (Qx \rightarrow Cx)$, $(Mx) (Qx \ \& \ Hx) \vdash (Mx) (Cx \ \& \ Hx)$.

(*B*1) turns out to be a *valid* form of argument, provable in the systems so far considered.

To assess the situation, it is necessary to set forth explicitly the grounds on which (*A*1) may be judged to be invalid. The argument would be this.* Suppose that Quakers form a very small proportion of Christians, and suppose that no Christians other than Quakers hate churches. And suppose the premisses of (*A*1) are true. The conclusion would nonetheless be false. The reason for this is that 'many' has *something* to do with proportions, in that the least number that can count as many of a much larger class is greater than the least number that can

---

*An elaboration of an argument put to me by John Skorupski.

count as many in a much smaller class. Hence two hundred, in this case, may be enough to count as many Quakers, but too few to count as many Christians. So, the argument runs, in informal argument the minimal manifold may change in size in the course of the same argument, whereas in W, S and T, the minimal manifold must remain the same size throughout any argument. This explains why $(B1)$ does not accurately represent $(A1)$.

However, although there is something in this argument, it is not quite accurate. For consider the simpler argument

(A2) There are many Quakers. All Quakers are Christians. Therefore there are many Christians.

(A2) seems to be perfectly valid, and accurately representable by

(B2) $(Mx)\,Qx,\;(\forall x)\,(Qx \to Cx) \vdash (Mx)\,Cx.$

(B2) is of course provable.

The intuitive difference between $(A1)$ and $(A2)$ points to the fact that the position in informal argument is more subtle than anything in this monograph has hitherto brought out. What seems to be the explanation is that there is an important logical difference between the subject-predicate sentence of the form 'Many Fs are Gs', and the simply quantified assertion 'there are many Fs', or 'there are many things which are both F and G'. And the difference is that whereas in the simply quantified assertion the size of the manifold is determined independently of the terms occurring, in the subject-predicate form, the *subject-term* must be taken into account in determining the size of the manifold. In the subject-predicate form, we have something analogous to the case of attributive adjectives.* In 'Martini is a poor violinist', the adjective 'poor' is not an *independent* determinant of the meaning. The sentence cannot be paraphrased as 'Martini is poor and Martini is a violinist', if the sentence is to be taken as meaning 'Martini is poor at playing the violin'. Similarly, in the subject-predicate cases of 'many', we cannot treat 'many' as having its meaning determined independently of the subject-term. Its force varies with the

*Thomas Baldwin drew my attention to this analogy.

subject-term. And *how* it varies is that the size of the minimal manifold varies. But the systems do not take account of this. In *their* interpretation it is assumed that the minimal manifold is fixed as the same for all terms occurring. Hence the systems are inadequate to represent variations determined by different subject-terms.

A striking way of bringing out the crucial effect of the subject-term is that 'Many Fs are Gs' is not generally equivalent to 'Many Gs are Fs'. For instance, 'Many Quakers are teetotallers' is not equivalent to 'Many teetotallers are Quakers', and 'Many pyramids are ancient buildings' is not equivalent to 'Many ancient buildings are pyramids'. But since '$(Mx)(Fx \& Gx)$' is equivalent to '$(Mx)(Gx \& Fx)$', formulae of these forms do not properly represent the subject-predicate sentences. Where subject-predicate sentences are involved in an argument, such a representation is admissible only where 'Many Fs are Gs' is equivalent to 'Many Gs are Fs', for every term occurring in the argument. But such a requirement limits the applicability of the logic rather drastically.

These limitations must be admitted, and overcome as far as possible. However, they do not show that there is anything *wrong* with the systems so far presented. They do not show that the systems cannot be applied at all, nor that they contain invalid principles. There *are* such applicable quantifiers as '(Mx)' and '(Nx)', and they are applicable with the intended interpretation. But they do *not* give an exhaustive account of the informal quantifiers.

A refinement of our present systems enables us to cope with this problem. As a preliminary to explaining the refinement, a little more must be said about the relation between 'Many Fs are Gs' and '$(Mx)(Fx \& Gx)$'. Suppose it is true that many Fs are Gs. The property F, we have agreed, is relevant in determining a suitable size of manifold. Let k be a number that represents that size of manifold. Then it will also be true that there are at least k things that are both F and G. Let us represent that proposition by $(M^k x)(Fx \& Gx)$. On the other hand, if k is a suitable number, then if '$(M^k x)(Fx \& Gx)$' is true, it will also be true that many Fs are Gs. In other words, it is quite permissible to let 'many Fs are Gs' be represented by a conjunctive proposition of the form '$(M^k x)(Fx \& Gx)$', so

F

long as k is a suitable number determined by the nature of F.

Now '$(M^kx)(Fx \ \& \ Gx)$' is equivalent to '$(M^kx)(Gx \ \& \ Fx)$'. But this latter formula may not be a good representation of 'Many Gs are Fs', since k may not be suitable as a number determined by the property G. Thus although we may represent 'Many Fs are Gs' by '$(M^kx)(Fx \ \& \ Gx)$', and transform the latter into '$(M^kx)(Gx \ \& \ Fx)$', we cannot always translate this last formula back as 'Many Gs are Fs'. Thus although we have an adequate symbolic rendering of 'Many Fs are Gs', as things stand we cannot say, given a symbolic rendering, what informal sentence it renders. This can be overcome by indexing the relevant predicate. Thus '$(M^kx)(F^kx \ \& \ Gx)$' is rendered 'Many Fs are Gs', and '$(M^jx)(Fx \ \& \ G^jx)$' is rendered 'Many Gs are Fs'. This indexing of *predicates*, however, will turn out to be irrelevant to the formal reasoning. It is relevant only to showing what informal sentences are being represented. The indexing system can be extended to more complex predicates, e.g. 'Many FHs are G' can become $(M^kx)((Fx \ \& \ Hx)^k \ \& \ (Gx))$. However, here an important point must be mentioned. It might be thought that the index of a complex predicate could be determined as a function of the indices of the atomic predicates from which it is made up. But this is not so. For consider a simple example: 'Many old Quakers say "thou" instead of "you".' One cannot determine the index of 'old Quakers' as a function of the indices of 'old' and 'Quakers', since the index of 'old Quakers' is determined by the number of old Quakers there are, and this cannot be calculated simply from the number of old people and the number of Quakers, since it depends on how far the class of old people overlaps with the class of Quakers, and this is an empirical matter. Thus just as the indices of atomic predicates have to be informally determined, on empirical grounds, so do indices of complex predicates, and these latter have to be determined independently of the indices of the atomic predicates from which the complexes are compounded.

Now consider the inference from 'Many Fs are Gs' to 'Many Gs are Fs'. This inference is valid only where the index of 'G' is no greater than the index of 'F'. Thus if and only if $j \leq k$, we have

$$(M^kx) \ (F^kx \ \& \ G^jx) \vdash (M^jx) \ (G^jx \ \& \ F^kx).$$

Similarly, the inference from 'Nearly all Fs are Gs' to 'Nearly all not Gs are not Fs' is valid only where the index of 'not G' is no *smaller* than the index of F. For example, consider

Nearly all businessmen are millionaires: so nearly all non-millionaires are not businessmen.

This seems to be valid. But compare

Nearly all the rich are sane; so nearly all the insane are not rich.

This seems to be *invalid*. The former can be expressed as

Not many businessmen are not millionaires; so not many non-millionaires are businessmen.

This also seems valid. The latter can be expressed as

Not many rich people are insane; so not many insane people are rich.

And this also seems invalid.

The contrapositives of these arguments are

Many non-millionaires are businessmen; so many businessmen are not millionaires.

Many insane people are rich; so many rich people are insane.

The former is valid, the latter invalid. And the reason is that given, for 'many', that the index of 'non-millionaires' is no smaller than the index of 'businessmen', so the first argument is valid. But the index of 'insane' *is* smaller than the index of 'rich', so the second is invalid.

Now this brings out a point about indices: all we need to know is their relative sizes, not their absolute numerical values. Consequently, we can treat the superscripts in the quantifiers simply as indicating a place in an ordering of the quantifiers, instead of indicating the exact size of a manifold. And this is preferable, since otherwise 'Many Fs are Gs' would collapse into a straight numerical statement 'There are n things that are both F and G' – a result we have hitherto avoided by systematic ambiguity. Henceforth, therefore, the indices will indicate only position in an ordering of the sizes of minimal manifolds. How this works in a formal logic we shall now see.

The system T^k

In addition to the notation of ordinary classical predicate calculus, the vocabulary of T^k includes first of all a finite sequence of penuniversal quantifier-signs

$$N^1, \ N^2, \ldots N^k, \ldots.$$

With every penuniversal quantifier-sign there is associated a positive integer, in such a way that if $j < k$, the number associated with N^j is strictly less than that associated with N^k. This association can be made in an infinite number of ways; the logic is systematically ambiguous between these associations.

Next, for each number n associated with some N^k, the vocabulary includes a list of $(2n-1)$-tuples of constants.

No further vocabulary is required.

The rules of T^k are the rules of predicate calculus, and in addition the following three sequences of rules.

NE^k. Given $(N^k v)A(v)$, one may derive $\bigvee A(g_i)$, resting on the same assumptions, where

(1) n is the number associated with N^k.

(2) $g_1, \ldots g_n$ are n constants all selected from the same $(2n-1)$-tuple.

NI^k. Given $\bigvee A(g_i)$, one may derive $(N^k v)A(v)$, resting on the same assumptions, subject to the restrictions that no g_i occurs either in $(N^k v)A(v)$, nor in any assumption upon which $\bigvee A(g_i)$ rests, and where

(1) n is the number associated with N^k.

(2) $g_1, \ldots g_n$ are constants all selected from the same $(2n-1)$-tuple.

k-change. Given $(N^k v)A(v)$, one may derive $(N^j v)A(v)$, provided that j is greater than k. The conclusion rests on the same assumptions as the premisses.

The interpretation of T^k consists of a domain of at least $2n-1$ elements, where n is the largest number associated with any penuniversal quantifier-sign, together with an assignment, for every n, to every constant in the sets of $(2n-1)$-tuples, in such a way that in any $(2n-1)$-tuple all the constants are assigned distinct elements. The interpretation of predicate letters is just as in predicate calculus.

The rule of truth for N^k is that, if n is the number associated with N^k, then $(N^k v)A(v)$ is true iff in every n-membered set from the domain there is at least one element with the property

A. It is fairly clear that the rules NE^k, NI^k and k-change are sound under this interpretation.

Corresponding rules for a sequence of quantifiers M^k can be derived. The rule of k-change for M^k is as follows: given $(M^k v)A(v)$, one may derive $(M^j v)A(v)$, provided that j is smaller than k.

The application of the system can now be more accurately discussed. In order to express informal arguments, all relevant predicates must be ordered with respect to the relative sizes of manifold involved, and given appropriate indices accordingly. For instance, if two of the relevant predicates are 'F' and 'G', we have to ask ourselves 'Would there have to be more Fs for there to be many Fs than there would Gs for there to be many Gs?' If so, the index of 'F' must be greater than that of 'G'. If there have to be fewer Fs than Gs to count as many, the index of F is smaller than that of G. Otherwise the index of the two predicates is the same. Sometimes there is a relevant predicate that is complex, such as 'fat bureaucrat'. Such complex predicates must be given indices separately from their simpler constituents.

Now, having given indices to predicates that provide a suitable ordering of the predicates with respect to the relative sizes of manifolds involved, wherever a sentence of the informal argument contains a quantified subject-term, either of the form 'Many Fs' or 'Nearly all Fs', the quantifier is represented by the formal counterpart with the same index as that of the subject-term. Thus, suppose one informal sentence is of the form 'Many Fs are Gs', and the index of 'F' is k and the index of 'G' is j. Then the sentence is represented by '$(M^k x)(F^k x$ & $G^j x)$'. The formal sentence thus shows the quantifier as keyed to the subject-term. 'Many Gs are Fs' becomes '$(M^j x)(G^j x$ & $F^k x)$'. Thus 'Many Fs are Gs' will be *equivalent* to 'Many Gs are Fs' only if 'F' and 'G' have the same index. On the other hand, there will always be one which follows from the other; if $k \geq j$, then '$(M^j x)(G^j x$ & $F^k x)$' follows from '$(M^k x)(F^k x$ & $G^j x)$', and if $k \leq j$, then the latter follows from the former.

It is clearly possible to formulate systems W^k and S^k, which relate to W and S respectively in the way that T^k is related to T. There are thus six systems (or strictly, sets of systems) for the logic of plurality. They differ from one another in the size of the presupposed domain of discourse, and in whether or not they take account of the attributivity of plurality-quantifiers. The larger the domain of discourse, the stronger the logic. The pattern into which the systems fall can be summarized in the following table.

	Domains:		
	Non-empty	n-membered	$(2n-1)$-membered
Non-attributive	W	S	T
Attributive	W^k	S^k	T^k

There is something to be said for favouring T^k above all the others. It has the advantage, through its attributivity, of representing informal quantifiers most closely and intuitively, and the other advantage, deriving from its strength, of validating the largest number of inferences within the field of plural propositions. The objection to it, that, since it demands a relatively large domain, it makes ontological claims of an order that logic should avoid, is not at all serious. In classical predicate logic, assuming a non-empty domain, rather than allowing the domain to be empty, is not justifiable on grounds other than the strength and workability of the resulting system. The justification of T^k seems to be entirely similar. The mere non-emptiness of a domain is in no way a privileged assumption. Aristotelian logic assumes that every *term* is non-empty, which is a strong restriction on its applicability, but that does not entail that there is anything *wrong* with it. A second objection to T^k, which applies also to all the attributive systems, is that to apply it the predicates must be ordered, and this can be done only on *empirical* grounds, not from considering the *meaning* alone of the predicates. This feature is, however, a merit rather than a defect. For in general, to the extent that a speaker is unclear what inferences can validly be drawn from his assertions, to that extent is he unclear what his assertions actually are. Granted the attributivity of plurality-quantifiers, it *will* be unclear what inferences are licensed unless the predicates are, at least implicitly, ordered in the way suggested. There are thus two alternatives; either attributive plurality assertions *are* inherently unclear, and have no determinate logic, or the predicates are implicitly ordered. The former alternative is false, since we can make sensible judgements of the validity and invalidity of attributive plurality inferences. So the second alternative must be accepted. If so, there can be no objection to a logic making explicit what is implicit in informal usage. This is indeed an important part of the point of formal logic.

In ordering the predicates, we are making our meaning more explicit and more determinate. That the ordering is done on the basis of empirical knowledge is important, but it is no objection. It brings out a philosophical point. Shared understanding of an attributive plurality proposition requires a common ordering of the predicates. A common ordering of the predicates requires similar empirical beliefs upon which to base it. Consequently, those who wish to use attributive plurality propositions can do so without fear of misunderstanding only in the context of a certain set of shared empirical beliefs. Since one who speaks generally hopes to be understood, there is here a condition on the suitable occasions of use of such propositions. Where occasions are suitable, the meaning will be clear, and the inferences licensed determinate.

Although T^k is the system most characteristic of plurality inferences, the other systems are not without interest. First, it is always of interest in formal pursuits to obtain results on a weaker basis where possible. For then we can determine whether a certain inference remains valid if the presupposition of the stronger system turns out to be false. W^k and S^k are of interest for this reason. W, S and T, the non-attributive systems, are of interest for another reason. For they show that many of the fundamental methods of reasoning with plurality-quantifiers are independent of their attributive nature. It was therefore possible to expound many of the essential ideas of the logic of plurality without taking attributivity into account. This procedure has both philosophical and expository value. Furthermore, the procedure shows that, besides the attributive quantifiers, there is also a set of quite coherent non-attributive plurality-quantifiers.

The fact that the universal and existential quantifiers are limiting cases of the penuniversal and plural quantifiers suggests the following hypothesis: that 'all' and 'some' are no less attributive than 'nearly all' and 'many', but this fact is masked by the circumstance that in the limiting case this attributivity has no logical effects. This hypothesis, which has a certain plausibility, would aid the construction of a unified theory of how quantifiers work. The purely logical backing to this idea is given by the present systems.

Appendix

There is little difficulty in constructing *intuitionistic* systems for the logic of plurality corresponding to each of the systems, W, S, T, W^k, S^k, T^k. The systems I have presented are all classical systems, and a logician who has scruples about certain of the classical principles for the universal and existential quantifiers will have precisely analogous scruples about analogous principles in the logic of plurality. For instance, just as an intuitionist denies the acceptability of a proof of $(\exists v)A(v)$ unless it provides some method whereby an example of an object with property A could in principle be produced, so he must deny the acceptability of a proof of $(Mv)A(v)$ unless it provides some method whereby n examples of objects with property A could in principle be produced. Consequently, just as

$$\sim(\forall v)\sim A(v) \vdash (\exists v)A(v)$$

is not an intuitionistic sequent of ordinary logic, the sequent-expression

$$\sim(Nv)\sim A(v) \vdash (Mv)A(v)$$

should not be a sequent of intuitionistic logic of plurality. Just as the universal and existential quantifiers are *not* interdefinable in intuitionistic logic, so the penuniversal and plural quantifiers are not interdefinable in intuitionistic logic of plurality.

Two main kinds of system have resulted from attempts to formalize the ideas of intuitionist mathematicians. One, the weaker, is known as the *minimal* logic of Kolmogorov-Johansson. The other is due originally to Heyting, and is simply called the *intuitionistic* logic.

The minimal predicate calculus is obtainable from the predicate calculus system of this book by the simple means of dropping one half of the law of double negation. We retain the rule that from A we can derive $\sim\sim A$, but abandon the converse, that from $\sim\sim A$ we can derive A. This has the result that the universal and existential quantifiers are no longer interdefinable, since the proof of

$$\sim(\forall v)\sim A(v) \vdash (\exists v)A(v)$$

requires the now abandoned half of the law of double negation. The rules for these quantifiers are thus now no longer non-independent, and all four are necessary. The intuitionistic logic is obtainable from the minimal logic by adding to it the rule that from an explicit contradiction an arbitrary well-formed formula may be derived. This does not restore the non-independence of the quantifier rules. Thus the minimal and intuitionistic predicate logics are obtainable from the classical system by changes in only the propositional calculus part of the system. The rules for quantifiers remain intact, but their import is different because of changes in the propositional bases.

Not surprisingly, these changes in the propositional calculus base are also all that need to be made to obtain also the minimal and intuitionistic logics of plurality, since the interdefinability of the penuniversal and plural quantifiers also depends on the dropped half of the law of double negation, and is not restored by the addition of the intuitionistic rule that from a contradiction anything may be derived. This is so whichever of the six systems I have presented one is dealing with. So in this way one obtains minimal and intuitionistic logics of plurality corresponding to each of W, S, T, W^k, S^k, T^k (the relevant rules ME and MI would have to be restored where I have omitted them from the classical systems, relying on their non-independence there). Minimal and intuitionistic semantics can then be given for these systems, using the ideas of this book to adapt the standard semantics for the ordinary minimal and intuitionistic predicate logics. The formulation of these semantics can become rather involved, and I forgo their presentation here.*

---

*For further information on intuitionist systems, see especially S. A. Kripke, 'Semantic analysis of intuitionistic logic I', in *Formal Systems and Recursive Functions*, (North Holland, 1965), or K. Schütte, *Vollständige Systeme Modaler und Intuitionistischer Logik* (Springer, 1968).

Each system is in fact, via systematic ambiguity, a de-numerable family of determinate systems. So we finish up with eighteen such denumerable families of logical systems, which may perhaps be enough to satisfy quite a variety of logical tastes.

Index